HOW THE NEW TECHNOLOGY WORKS

HOW THE NEW TECHNOLOGY WORKS
A GUIDE TO HIGH-TECH CONCEPTS

by
Robert J. Cone

Revised and Updated
by
Patricia Barnes-Svarney

ORYX PRESS
1998

The rare Arabian Oryx is believed to have inspired the myth of the unicorn. This desert antelope became virtually extinct in the early 1960s. At that time several groups of international conservationists arranged to have 9 animals sent to the Phoenix Zoo to be the nucleus of a captive breeding herd. Today the Oryx population is over 1,000, and over 500 have been returned to the Middle East.

The first edition of this book was titled *Key to High-Tech: A User-Friendly Guide to the New Technology*.

All illustrations in this edition drawn by Gregg Myers.

© 1998 by The Oryx Press
4041 North Central at Indian School Road
Phoenix, Arizona 85012-3397

Published simultaneously in Canada
Printed and Bound in the United States of America

∞ The paper used in this publication meets the minimum requirements of American National Standard for Information Science—Permanence of Paper for Printed Library Materials, ANSI Z39.48, 1984.

Library of Congress Cataloging-in-Publication Data

Cone, Robert J.
    How the new technology works: a guide to high-tech concepts/by
Robert J. Cone: revised and updated by Patricia Barnes-Svarney.
    Includes bibliographical references and index.
    ISBN 1-57356-138-X (alk. paper)
    1. High technology—Handbooks, manuals, etc.   I. Barnes-Svarney,
Patricia L.   II. Title.
T49.C66   1998
600—dc21   98-9169
                                                              CIP

Contents

Preface

The high-tech revolution is a broad movement founded on important technological advances. It embraces significant issues of commercial, political, and cultural practice and style.

The term "high-tech," in current popular usage, refers to that segment of advanced technology that directly affects the day-to-day lives of people in their homes, in their cars, on their jobs, and in their doctors' offices. This is the sense in which the term is used in these pages.

The key words of the high-tech world (chip, laser, etc.) are quite familiar to the public. Articles about new high-tech capabilities and new applications in commerce, medicine, and the home appear more and more frequently in the media, but the information is often fragmentary and unfocused. *How the New Technology Works* addresses this problem. It avoids introducing engineering details and operational minutiae that are not needed by the intelligent "outsider" who is seeking a basic understanding of the general principles upon which the device or mechanism in question operates.

While most people are intensely interested in advanced technical devices such as robots and lasers that are now affecting, or may soon affect, their daily lives, they are not particularly interested in laboratory spectrometers or scanning electron microscopes, which are perceived as remote from their personal concerns. A careful survey of media coverage of technology, in addition to extensive discussions with members of the scientific and commercial communities, formed the basis of selection for the particular topics included in this book.

This new, updated version of *How the New Technology Works* provides concise, lucid explanations of the selected topics in a systematic presentation. It is divided into three main sections. The first section covers 17 major high-tech devices and systems; section 2 provides a nontechnical overview of electromagnetic, communications, and measurement concepts; and section 3 covers 20 secondary devices and systems. The discussion of concepts and theory has been deferred to section 2 because most people have a more immediate interest in the capabilities and operating principles of the new devices than they do in underlying theory. It is hoped that the knowledge acquired by reading section 1 will spur the reader's desire to seek a greater understanding of these underlying concepts as presented in section 2. In some cases, the reader may be asked to refer to articles in section 2 for more detailed explanations of scientific concepts key to that device's operation.

Each major topic covered in section 1 includes a concise description of the system or device in question, a list of applications in which it is being used, a comprehensive background discussion of how it was developed and how it works, and a projection of how it may develop in the future.

The reader can obtain a quick grasp of what the device or system is all about by reading only the basic description; reading the list of applications will provide an even better understanding; reading the background narrative as well will provide comprehensive knowledge of the device or system.

Although this book serves as a very useful reference source, *How the New Technology Works* could be read in consecutive order in its entirety. Some readers, however, may prefer to read the section on theory—section 2—before going to the first section on devices and systems.

This is the third edition of this book. Previously titled *Key to High-Tech* (the first edition), *How the New Technology Works* is completely updated and expanded, and illustrations have been provided to help clarify technical points. A full section on micromachining and nanotechnology has been added to the text; in addition, the previous secton on genetic engineering has been expanded into a section covering all of biotechnology. Section 3 also includes several new and updated technologies found in today's fast-paced world, including the Global Positioning System (GPS), microscopes, and planetary spacecraft.

Robert Cone wishes to extend his thanks to Wilbur Bull, Jr., John S. Hayward, and Arthur A. Whitfield who read all or part of the manuscript, to Roland S. Gohlke who read the section on superconductivity, and to Paul E. Weiss, MD, and Theodore F. Van Zandt, MD, who read the section on medical imaging. Patricia Barnes-Svarney wishes to thank Thomas E. Svarney for his research and editing work in all parts of the manuscript.

Note: Technical words and expressions are defined the first time they appear in a section. When a technical term appears with a plus sign (+) next to it, a detailed explanation of that term may be found elsewhere in the text. Referring to the index will provide the required page numbers. In cases where multiple page numbers occur in the index entry, the pages for the primary definition will be in boldface type.

HOW THE NEW TECHNOLOGY WORKS

Section One

Seventeen Major High-Tech Topics

This section provides a comprehensive review of 17 major systems and devices that have brought the new technology into the homes, offices, factories, and hospitals of the world's advanced nations.

Readers unfamiliar with basic scientific principles may prefer to read section 2 first, before going on to the devices and systems described in the other sections.

Artificial Intelligence

BASIC DESCRIPTION

Artificial intelligence (AI) is the ability of a computer (or other machine) to reason and reach conclusions through "thought" processes similar to those employed by human beings. Currently, no machines have achieved this capability in its truest sense. Such a machine could, however, perform useful actions that could not be predicted with certainty without it. For example, a true AI machine, if assigned the task of determining whether the Earth's climate was beginning to undergo a significant transformation, could (without human intervention) select appropriate criteria from the data available, develop and operate a program model projecting the Earth's climate over the next 10 or 20 years, and reach a reasonable conclusion that would not be predetermined by its programming rules and hardware constraints.

Note: The term "artificial intelligence" is also used, particularly in commercial circles, to refer to many existing software programs and hardware devices that perform relatively sophisticated functions but are not "thinking" or demonstrating originality in any fundamental sense. These programs and devices are operating from rigidly predetermined instruction sets, and their results, however complex the process by which they are attained, are predictable.

REPRESENTATIVE APPLICATIONS

Commercial AI

In the limited sense in which the term AI is being used commercially at the present time, a number of ongoing applications are achieving some measure of success:

- Language perception (foreign language translation; natural language interaction [the ability to interact with the computer using ordinary words at conversational speeds])

- Perception techniques (3-D modeling; autonomous navigation; expert systems; machine vision [+])
- Reasoning (expert systems; fuzzy logic)
- Robot (+) control
- Logic games (chess playing)

Theoretical AI

Artificial intelligence, as understood in the fundamental sense (intelligence that truly emulates human reasoning), does not yet exist. In the opinion of some experts, it will never exist. If achieved, however, potential applications for it would be almost limitless. Immune to fatigue, computational errors, and moods, true AI devices could complete complex, sensitive activities now performed exclusively by human beings.

In military applications, for instance, the ability to use intelligent, expendable devices in hazardous situations would confer a tremendous advantage on the side that possessed them. For example, an intelligent robotic device capable of performing many of the duties of a human soldier could remain active within an area where the enemy might use poison gas or biological weapons. The U.S. Department of Defense regards the development of AI as one of its most critical objectives.

Theoretical AI is also referred to as "strong" AI.

BACKGROUND

Charles Babbage, a British mathematician and physicist (1791-1871), is generally credited with developing the first conceptual blueprint of a digital computer. Unfortunately, his brilliant design outstripped the technology of the day. The complicated assembly of gears and wheels that would have been required to make it a reality presented too great an engineering problem.

The "Analytical Engine," as Babbage called it, never really became a practical device.

Even in those early days of the computer, however, people were concerned with the future implications of "thinking machines." Ada Lovelace, a gifted scientific enthusiast and Babbage's friend, tried to encourage a realistic attitude on the subject when she wrote:

> The analytical machine has no pretensions whatever to originate anything. It can do whatever we know how to order it to perform.

So far, Lady Lovelace's assertion has held up. Although computers are now performing tasks of bewildering complexity, they are still doing only what we are able to tell them to do. But the idea of a "thinking machine" remains a compelling issue for society, and the topic has generated enormous amounts of discussion and commentary, particularly since the phrase "artificial intelligence" entered the language more than a quarter of a century ago.

AI continues to be the focal point for reflections and observations from those representing disciplines as diverse as mathematics, philosophy, natural sciences, sociology, and history. If nothing else, it has forced society to examine closely what is meant by such words as "intelligence," "creativity," "common sense," and "consciousness."

TYPES OF ARTIFICIAL INTELLIGENCE

As indicated above, there appear to be two major groupings working under the broad banner of artificial intelligence. One consists of those scientists and engineers who concentrate on developing practical devices and systems to address the ongoing needs of industry, medicine, technology, and research, and the other consists of those who are dedicated to fundamental theoretical research.

Commercial AI

Of the practical, ongoing applications of AI, expert systems have been very successful. An expert system is a specialized computer program that can analyze complex situations within a particular area of expertise, reach useful conclusions about those situations, and offer recommendations for appropriate action. Some expert systems take the place of human experts, whereas others are designed to aid the expert. They are usually created by computer specialists in collaboration with experts in a particular field. Expert systems are made up of two primary components: a knowledge base containing extensive information about the particular field, and an analytical system that evaluates incoming data by referring to the knowledge base. Many successful expert systems have been developed in the fields of science and engineering, where they relieve the human expert of much tedious calculation and analysis by analyzing such things as chemical structures, and diagnosing human illness. They have also made a significant contribution in other areas such as developing industrial plant schedules, scheduling routes for delivery vehicles, and, in general, relieving human experts of the burdensome, but essential, skilled drudgery of certain procedures.

In addition, commercial AI studies have also led to the development of "fuzzy logic," a process that offers a way to characterize subjective ideas. Contrary to the usual Boolean logic used in computer programs, which uses simple "yes" and "no" answers, fuzzy logic (in a way, a kind of Boolean logic) computationally solves problems that are filled with ambiguous data or "partial truths" (for a simple example, is the room too cold, too hot, or just right?). Simply put, fuzzy logic recognizes the probabilities of truthfulness and falsehood, thus making it useful in expert systems and other artificial intelligence applications. First named in the 1960s by electrical engineer Lotfi Zadeh, modern fuzzy logic applications are now being explored for such fields as engineering computing, economics and finance, and criminology, especially in expert systems. In addition, elementary forms of fuzzy logic are being used in many common appliances, such as toaster ovens, in which computers monitor, for example, the cooking process; the zoom control in hand-held cameras; and in the automatic transmissions of automobile engines. Another simple form of fuzzy logic is found in a computer software's spell-checker, which suggests a list of probable words to replace a misspelled one.

On the factory floor, AI has been only a moderate success. There has been much discussion of computer-controlled factories and computer-integrated systems, but relatively little has actually been accomplished, as most industries have found that the human element is still needed.

In the area of foreign language translation, programs can do some translation but they still require extensive human intervention. Progress on natural language interaction (the ability to interact with the computer using ordinary words at conversational speeds) was limited in the past. Currently, there are many computer software programs that allow the user to dictate or give simple commands to a computer (called voice recognition software), but they use a great deal of computer memory, and many are slow in interacting with the user. Several recent software breakthroughs will allow faster interaction between the computer and the user, especially when neural networks are used (see ahead).

One ongoing application that has generated a great deal of interest in the media and with the public is the chess-playing computer. While it is not clear what the real-world utility of such a device is, researchers feel it is an aid in understanding other complex systems, and it has the advantage of being easy to test. (You simply have it play against humans or other computer programs.)

Many computers have been tested against chess champions recently. Several years ago, Deep Thought, a program developed at Carnegie Mellon University, defeated a number of chess masters before it was roundly trounced by the world champion in a two-game match. However impressive its accomplishments, Deep Thought was probably a disappointment to AI theoreticians. Its primary developer, Feng-hsiung Hsu of Carnegie Mellon, readily admits that his program did not simulate, or in any way resemble, the thinking processes of a human grand master. Quite to the contrary, it reached its conclusions by "brute force," examining around 34 million positions a minute and evaluating each one on a predetermined quantitative scale. Human experts, on the other hand, look ahead to a relatively small number of positions and select those few that they will analyze exhaustively by a process of intuition, common sense, and the kind of ability to recognize patterns that comes with long experience. (Actually Deep Thought did use some elements of selective search, but its primary strength was in its ability to examine millions of positions.) The program, despite its power, could not make independent judgments based on experience or human-like reasoning.

Another supercomputer (+) specially programmed to take on the world's best human chess players was Deep Blue. The software and hardware for Deep Blue were developed by the IBM computer science team of Chung-Jen Tan, Feng-hsiung Hsu, Murray Campbell, A. Joseph Hoane, Jr., and Gershon Brody and assisted by former U.S. Chess Champion Grand Master Joel Benjamin. Deep Blue played against world champion chess player Garry Kasparov in 1996 and lost. But in 1997, after further refinements by programmers, Deep Blue beat Kasparov, the first computer to win such a tournament. Again, the computer could not make independent judgments based on human-like reasoning. It chose its moves from a database based on over 600,000 master games and could analyze 400 million chess positions per second.

Theoretical AI

The question of whether or not computers can become truly intelligent is hindered by the lack of a generally accepted definition of the word "intelligence" within the AI community. Rather than defining the word, the tendency is to suggest tests for determining whether it is present in a computer program or system.

One of the landmark tests was offered by Alan Turing, a distinguished British mathematician and computer pioneer. He suggested that a computer be placed behind a curtain and that people be allowed to "talk" to it via a keyboard or similar device. If, after allowing people to communicate freely on any subject through the curtain, the testers could not determine whether it was a person or a machine in there, the machine would be deemed to be intelligent. The curtain would be necessary because we tend to judge things by their appearance, i.e., a human is automatically assumed to be intelligent, and something that looks like a machine is automatically judged to be nonintelligent.

Note: The first Turing test competition took place on November 8, 1991, at the Computer Museum in Boston, Massachusetts. A selected group of computers were subjected to questioning by a varied group of individuals under the conditions laid out by Turing. The Loebner Prize, as it is called, is now an annual event, in which a $2,000 award and bronze metal are awarded to the most human computer. Dr. Loebner pledged a Grand Prize of $100,000 for the first computer whose responses were indistinguishable from a human's.

The Turing test, however, says nothing about what is going on inside the machine. Does it matter? Should the judgment be made purely on the basis of performance? *Merriam Webster's Collegiate Dictionary* says that intelligence is "the ability to learn or understand or to deal with new or trying situations." Would the Turing test show whether a machine possessed such abilities?

In the case of chess-playing computers, we have seen that, in reality, they understand nothing at all about the game of chess but simply generate rule-based arrangements of symbols (pieces on the chess board) for so many moves ahead, compare the arrangements on a quantitative scale, and generate their moves on a purely mechanical basis.

Professor John R. Searle of the University of California at Berkeley shows even more clearly how intelligence can be simulated with the example of the Chinese room, which is paraphrased below:

An English-speaking person, with no knowledge of Chinese, is placed within a closed room, given a rule book in English, and also given baskets full of individual Chinese characters. The rule book identifies the characters only by their shape and gives no clue as to their meaning.

A native Chinese speaker, outside the room, asks a question by inserting a set of Chinese characters through a slot into the closed room. The person in the room looks up the set of characters in a rule book and is instructed to remove certain other characters from the baskets and pass them back out in a certain order through the slot. If the question in Chinese was, for example, "what are your favorite colors?" and the characters passed outside meant "red and blue," the questioner would feel that an intelligent answer had been given. A continuing dialogue might well convince the native Chinese speaker on the outside that he was dealing with an intelligent Chinese speaker on the other side of the wall.

In this situation, the person in the closed room is like the central processor of a computer, the rule book is like a program, and the individual's actions in carrying out the rules are purely mechanical. The individual in the closed room has carried on a conversation in Chinese without knowing a single character of that language or understanding a single word. The human has dutifully done what computers have always done, i.e., manipulated symbols according to rules. The weakness in the system is, of course, if the speaker asks a question not covered by the rule book, the whole process breaks down.

Following this approach, it is difficult to visualize a promising path leading to genuine artificial intelligence. The difficulty is that, while computers do a good job with rule-based logical systems like chess, they have a much harder time dealing with the fuzziness and disorder of everyday real-world situations. Just distinguishing between a book and a box lying near each other on a table is a very tough job for a computer equipped with a visual sensor.

In light of these problems, some investigators have taken a different route and are attempting to develop computer structures that resemble the human brain.

Neural Networks

The communications network in the human brain is far from being fully understood, but it is known that signals, in the form of electric pulses, are generated by varying relationships among charged biochemicals that are usually present in the form of ions (charged atoms).

The basic component of the brain's network is the neuron, which consists of a cell structure plus dendrites and axons. There are perhaps a hundred billion neurons in the brain. The cell body of a neuron is spherical in shape and contains a nucleus. The dendrites (which project out from neurons) are short and thick at the base and branch out like a tree. They are the information input components of the neuron. Axons are thread-like structures that connect the cell body to other neurons. They are the information output channels of the system.

A neuron may connect with as many as 10,000 other neurons. At the point of connection, which is known as the synapse, a "decision" is made whether or not to accept an incoming pulse. Thus, the synapse serves as a gate or switch.

Depending on the pulses that are passing through a system, their strength, and in what combinations they are received by a neuron, the neuron either "fires" or is inhibited. When the neuron fires, it sends pulses to other cells; when inhibited, it doesn't transmit the pulse. When these kinds of purposeful firings and nonfirings are coordinated over many millions of neurons within the brain's network, some significant event (or events) may take place in the brain.

For example, if one is driving a car on a dark road and suddenly sees a brightly lit neon sign ahead, one's eyes sense the sign and begin the process of transmitting signals (electrical pulses) across the nerve system to the brain's network, where an image forms in a tiny fraction of a second. Almost at the same instant, in ways not fully understood, memory cells are referenced in the mind so that the letters in the sign can be understood. During the fraction of a second or so within which this image is formed and identified, millions of neurons have fired, others have been inhibited, and the net result of all this electrochemical activity is that the driver can make the appropriate decision with respect to the sign.

AI researchers who are pursuing the brain-model approach have constructed electronic circuits that simulate, in a limited way, the brain's network.

The most striking difference between this method and a conventional (digital) computer is that each node (or neuron) is actually a processor. Where the conventional computer has a single processor (referred to as the CPU or central processing unit), or a few processors, the neural network (NN) computers will have thousands, or millions. Where the CPU of a conventional computer is a high-capability processor, the thousands of small processors in the NN will have only a very limited capability.

The basic objective of the NN computer is to provide a mechanism that, like the human brain, can learn through positive feedback which answers are right and which are wrong. The NN system does this by providing quantitative weightings to the signal connections among neurons. It strengthens the connections for a particular pattern if the system performs well and weakens them if it doesn't. As weightings are developed for various patterns, the computer performs with increasing accuracy. In effect, it has learned.

The weighting mechanism is provided by resistors, which can be adjusted up or down to weaken or strengthen the pulse. Since the NN will have thousands or millions of neurons, the failure or loss of a few of them will not significantly affect the operation of the device. Some pilot systems, using these principles, have been built and show considerable promise.

Scientists have developed several current applications of artificial NN. For example, NN is used in the collection and analysis of remotely sensed data of the Earth for environmental analysis. Many craft take a multitude of data, such as visible and other electromagnetic readings, and temperature and humidity data. These data are then easily classified, and the self-organizing NN can cluster the imagery data, for example, to recognize certain types of clouds; this is unlike expert systems, which are usually decision-making tools that can quickly analyze a larger amount of data faster—both methods are the closest so far to how the human brain works. In addition, NN has also been used to model mind-brain processes, another data and complex intensive study.

THE FUTURE

AI still has a long way to go. Deep Blue has defeated a human world chess champion, and now researchers are hopeful that it can be done again. But again, the chess-playing computer has nowhere near the type of intelligence humans possess.

Neural networking (which is also loosely referred to as connectionism) has gained a number of significant converts in recent years, some from the ranks of those who have been particularly skeptical about the genuine potential of AI. For example, with the new software technology, some researchers are predicting that NN will be combined with the computing power of the desktop PC, with enough power to predict the stock markets. Those who are skeptical about AI feel that if it is ever to make real progress it will have to leave the logical, rules-based, symbolic world of the digital computer and enter the fuzzier domain (including the use of fuzzy logic) of million-neuron networks interconnected in a vast feedback learning system.

Some traditional AI people, still convinced that the road to AI is through rules-based logic systems, argue that the human brain represents only one model of intelligence and possibly not the best one. They point out that those early pioneers who tried to achieve human flight by simulating the wing flapping of birds failed miserably. In their eyes, the speed and power of a modern jet make a strong argument for looking beyond nature's models. In fact, many labs are taking a different approach to AI, investigating whether swarms of semi-smart robots (+) can generate a collective intelligence that is greater than the sum of its parts.

FURTHER READING

Books

The Age of Intelligent Machines. Raymond Kurzweil. Cambridge, MA: MIT Press, 1992 (rep. edition).
    An authoritative presentation of AI issues.
After Thought: The Computer Challenge to Human Intelligence. James Bailey. New York: Basic Books, 1996.
    A fascinating look into how computers will one day think in their own way, rather than mimicking human thought.
Artificial Intelligence. Patrick Henry Winston. Reading, MA: Addison-Wesley Pub. Co., 1992.
    A revision of the single bestselling introduction to AI ever published.
The Emperor's New Mind. Roger Penrose. Oxford, England: Oxford University Press, 1989.
    A penetrating look at the potential of AI by a distinguished British physicist.
Hal's Legacy: 2001's Computer as Dream and Reality. David G. Stork (ed.). Cambridge, MA: MIT Press, 1997.
    Explores the technologies of creating a HAL-like computer (from the science fiction story *2001: A Space Odyssey* by A.C. Clarke).

Articles

"Artificial Genius," by M.A. Boden, *Discover*, October 1996, v. 17, 104-07.
    Recent findings and projections about AI.
"Expert System Software Aims to Speed Maintenance." P. Proctor. *Aviation Week & Space Technology*, June 10, 1996, 53.
    How expert system software is used in aviation maintenance.
"HAL et al.: How Smart is Artificial Intelligence?" B. Stone, *Newsweek*, March 1997, v. 129, 10.
    The search for AI.
"Is the Brain's Mind a Computer?" by John R. Searle and "Could a Machine Think?" by Paul M. Churchland and Patricia Smith Churchland both appear in the January 1990 issue of *Scientific American*, 26-37.
    These companion articles examine AI from differing perspectives.

Biotechnology and Genetic Engineering

BASIC DESCRIPTION

Biotechnology is the use of microorganisms, both plant and animal cells, to produce or improve useful materials, such as food, medicine, and chemicals. Biotechnology has its roots in ancient history: Vinegar, beer, wine, and cheese are all produced by using biological processes that occur within living cells—and are all considered to be biotechnological processes. Today, biotechnology is also used to produce detergents and other chemicals familiar to our everyday lives.

Genetic engineering, considered one of the most important studies under the umbrella of biotechnology, is the process of changing some characteristic of a living organism by introducing a foreign, or altered, gene into its cells. (Genes are the basic units that control the transmission of inheritable traits from parent to offspring.) Cells of bacteria colonies can be altered by this process so that they will produce specific proteins or hormones that can be used for health treatment in human beings or animals, or for other useful purposes. Recently, altered genes have been introduced directly into human beings suffering from specific diseases, in the hope of overcoming or controlling those diseases.

REPRESENTATIVE APPLICATIONS

Agriculture

Still relatively new, genetically engineered crops may eventually mean far larger yields per unit of farm acreage. Using these techniques, scientists have already adjusted the nutritional content of some plants and have produced disease- and herbicide-resistant agricultural products. In the near future, they hope to produce plants that are more disease resistant; more resistant to drought and extremes in heat or cold; and that will grow in poorer (such as salty) soils, thus allowing plants to grow in less than ideal conditions. For example, genetically altered bacterial sprays, aimed at such objectives as the prevention of crop damage from freezing air temperatures, have also been developed.

Improving livestock and other domestic animals (including health and performance) by conventional breeding techniques is a traditional and accepted procedure, but it is difficult and slow to carry out in practice. Genetic engineering makes it possible to speed up these breeding processes dramatically, giving rise to the possibility of radical improvements and changes in the basic traits of the animals. Two of the many products that will ensure greater yield per animal and better quality animals are genetically engineered animal vaccines for such diseases as footrot and genetically engineered microbes (microorganisms) that produce animal growth hormone. The use of biotechnology to develop new drugs and vaccines for animal health is also growing. Just as promising is the mapping of the genomes (natural "blueprints") of food animals, such as cattle, sheep, and poultry. By controlling the genetic basis of many of the animals' traits, scientists hope for such improvements as better growth rates and litter size.

Marine Biotechnology and Aquaculture

Biotechnology is also used in association with marine environments. New compounds are being developed from marine organisms to fight inflammatory diseases and cancer. Genetically engineered fish are more resistant to disease and faster growing.

Industry

A number of genetically engineered bacteria show high potential for industrial applications. Examples: bacteria that can convert old newspapers and wood chips into sugar; bacteria that can absorb oil and speed up oil spill clean-ups; and genetically engineered yeast that can accelerate the fermentation process in wine production. Still other naturally occurring and genetically altered bacteria are used in the cleanup of harmful toxicants in the environment.

Medicine

Biotechnology is already used in the medical arena. For example, a bacterium produces the vitamin B12, and a mold produces B2, both selectively bred so that they manufacture a thousand times more of the vitamin than they would naturally. Many new drugs are also the result of biotechnology. Drugs can kill pathogens (disease-causing agents) outright or stop their growth by inactivating their enzymes. For example, the antibiotic penicillin binds to an enzyme on the cell wall of an invading bacterium, not allowing the cell enzyme to finish assembling the cell wall, thus killing the bacterium. Scientists are now using computer-modeling techniques to determine how diseases affect the body. Genetic medicines, those made to attack viral and genetic diseases, are also being explored using computer modeling.

In addition to the possibility of genetically altering human genes to eliminate certain inheritable diseases (techniques that are currently in the experimental stage), genetic engineering is now able to produce, economically and in quantity, certain vital substances that are lacking in the metabolism of abnormal individuals. Although the periodic injection of these substances into the individual's bloodstream, or other appropriate site, will not cure the disease (as a gene alteration might), it will alleviate or eliminate symptoms and in many cases allow the individual to lead a normal life.

Substances such as insulin, interferon, human growth hormone, vitamins, vaccines, and antibiotics either have been, or are on the verge of being, produced through genetic engineering techniques. Accumulating useful quantities of these substances, either from natural sources or from chemical synthesis, is a slow and expensive process. Genetic engineering makes it possible to produce and harvest such rare substances by growing them in colonies of altered cells, usually those of bacteria. It is in this area that most current commercial effort is focused and here that near-term profitability is most likely to be realized.

Waste Management

Biotechnology has also been used to handle human waste problems. Sewage processing plants around the world use microbes to consume the wide range of solid waste materials. Bacteria have also been used for larger waste management problems, such as genetically altered bacteria that feed on oil slicks; additional microorganisms are being tested to break down different hazardous wastes in the environment, such as pesticides, herbicides, and other chemical wastes.

Research

The ability to directly modify gene structure and function has contributed to advances in cancer research and other critical areas of medical research. Genetic engineering has become an essential technique in biomedical research laboratories.

Biotechnology techniques discovered long ago are now being explored as possible sources of new fuels. For example, researchers have long known that microbes produce alcohol from sugar; now they are working to see if this knowledge can be used to produce a renewable source of fuel, one that is less polluting than conventional fuels.

Genetic engineering has also become important in the study of animal cloning, in which an egg cell from an adult species produces a (somewhat) exact copy of the original. This type of genetic engineering is controversial and has led to numerous ethical and moral discussions, especially regarding the possible cloning of humans.

General

As with many of the high-tech topics discussed in these pages, the most revolutionary achievements of biotechnology and genetic engineering lie in the future.

It should be noted that since so many genetic engineering procedures involve the introduction of foreign genes into colonies of bacteria, there has been concern about the possible health and ecological consequences of these techniques. For instance, one concern is the possibility of the creation of new diseases to which human beings would have no natural immunity. Another is the possible introduction into the ecology of unpredictable new forms of elementary life that might upset the balance of nature.

These concerns have been somewhat relieved by the passage of safety legislation governing laboratory security, as well as by the fact that experience to date has shown that these harmful consequences are not high-probability occurrences.

BIOTECHNOLOGY BACKGROUND

Humans have practiced biotechnology for more than 8,000 years, especially in making bread, wine, and cheese. Dough would rise with the use of yeasts; fruit juice would become alcoholic to produce wine; and milk would sour to produce cheese.

In each of these circumstances—and although our ancestors didn't know the reason—microorganisms were responsible for the fermentation process, essentially a form of biotechnology. Certain characteristics of various foods, such as flavor and texture, are partly due to the activity of microorganisms including bacteria, fungi, and yeast. For example, yeasts are important to the food industry, as they ferment sugar to produce the alcohol in wine and beer, and are responsible for the bubbles of gas to leaven bread. Bacteria are also used in such foods as yogurt and cheese (the bacteria ferment the milk or cream), cured ham, and pickled cucumbers. Fermentation by fungi is also a form of biotechnology, and is used in such foods as tempeh, miso, and soy sauce.

Our ancestors also practiced crude forms of biotechnology when working with plants and animals, picking and choosing the best crops and animals for further use. Stone-Age farmers began planting seeds (the beginning of the agricultural era) of wild plants. Seeds from the most productive plants were saved, then used for the next year's plantings. Animals for food, fiber, and work were also improved in the same way.

GENETIC ENGINEERING BACKGROUND

In 1866, an Austrian monk, Gregor Johann Mendel (1822-1884), published the results of his cross-breeding experiments with garden peas. The study appeared in an obscure provincial journal, where it remained largely unnoticed until the turn of the century. In spite of the limited resources that he had to work with, and in spite of the fact that he had to base his conclusions entirely on external observations and could only guess at the internal processes that enabled his garden plants to pass on physical traits to their offspring, Mendel, in his article, had laid the foundation for understanding the basic laws of heredity for all forms of life.

We now know that living creatures are organized as collections of cells. The smallest living entities, such as bacteria and protozoans, exist as single cells; humans are made up of about 50 trillion cells. Probably the most extraordinary characteristics of living creatures are their ability to reproduce themselves and to attune their behavior to the twists and turns of their environment.

On the cellular level, both reproduction and survival require the precise performance of staggeringly complex sequences of internal operations. Somehow the necessary series of events must take place in the correct order, and within the necessary time frame, to mesh with other events underway throughout the living body. Where these events involve the formation of bodily substances, such substances must up-

hold rigid quality standards or they will be useless or even harmful.

If you cut your finger, for example, your body's system will begin immediately to take steps to repair the damaged area. If a bacterial cell is invaded by a toxic agent, it will take rapid action, sometimes successfully, sometimes not, to break down the invader or otherwise render it harmless.

These actions take place in both intelligent and nonintelligent creatures, because they are generally independent of any conscious mental effort. But they take place with great precision and purpose, and relatively few mistakes are made. The body will cover the damaged area of your finger with new skin, not with feathers. A cow will give birth to a calf, not to a bear cub. Each of these body systems behaves in a reasonable and consistent manner. It is logical to conclude then, as Mendel did, that, somewhere within their structure, living bodies contain an inherited set of rules that enables them to do the things necessary to reproduce and to survive.

Mendel's work was comprehensive and served as the basis for many specific genetic laws, but perhaps his most important single conclusion was that a gene for each inherited trait is given by each parent to each individual offspring. In the case of simple life forms like bacteria, a single identical set of genes is passed on to the offspring cells.

In 1944, a researcher named Oswald Avery, and his colleagues, C.M. MacLeod and M. McCarty, working at the Rockefeller Institute in New York, discovered that the instruction sets (now called genes) that caused Mendel's plants to behave as they did were present in, and transmitted by, a substance that is present in all life forms, namely, deoxyribonucleic acid (commonly shortened to DNA).

We now know that every living creature carries within its cells, or cell, a tiny concentration of DNA that carries the genes required for that creature's survival and reproduction. In the case of higher order life forms, the individual will have two sets of genes, one contributed by each parent.

DNA is made up of long strings of bases. (A base is an organic ring compound containing nitrogen, hydrogen, oxygen, and carbon.) Only four bases are present in DNA: adenine, guanine, cytosine, and thymine. Genetic researchers tend to use the first letters of these bases—A, G, C, and T—to identify them. This system seems quite appropriate, because these compounds make up a code or alphabet to spell out all the information necessary for the survival and reproduction of a living individual. This alphabet, with a few exceptions, is universal. It will spell out the same information in all life forms, whether complex like mammals, or elementary like bacteria. It has "punctuation" in it, i.e., start-stop codes and various other signals.

DNA in bacteria may contain about 4,200,000 bases; a human being's DNA will contain about 6,600,000,000 bases in each cell. (The four bases A, G, C, and T are used over and over again in varying combinations to make up these huge numbers.)

The DNA structure is made up of pairs of bases; however, the pairs are not arranged in all possible combinations. A is always paired with T, and G with C. This set pairing is a physical limitation. The bases are shaped in such a way that they can only physically interlock with each other in those specific pairings. These paired structures form the famous "double helix" that was discovered by Francis Crick, a British molecular biologist, and James Watson, an American biochemist, in 1953.

Any continuous stretch of base pairs along the DNA segment that can specify how to make a particular bodily substance is a gene. A single human gene can vary in length from under 1,000 base pairs to as many as two million. The substance that the gene specifies may define some recognizable trait of the individual; for example, in human beings it may result in red hair instead of black hair.

In the DNA structure received from either human parent, there may be as many as 50,000 to 100,000 genes. DNA within the cell is embedded in threadlike structures of tissue called chromosomes. A normal human cell has 46 chromosomes, 23 having been inherited from each parent. One of the primary tasks of genetic researchers is determining what genes are on what chromosomes.

Among the important functions of genes is that of providing the necessary information for the formation of proteins. Proteins are absolutely essential to all life forms and are present in all living cells. They serve as structural matter—forming skin, feathers, and cell coats and walls; they serve as transport mechanisms like hemoglobin, which moves oxygen through the blood; and they serve as regulatory agents such as hormones, which control some phases of metabolism. Some proteins are poisonous; snake venoms, for example, are proteins.

All proteins are vitally important, but among the most interesting and versatile are the enzymes. Enzymes are catalysts, i.e., they are substances that accelerate and facilitate reactions among other substances, without being changed themselves. Thus, an enzyme may help the merging of two compounds into a single new compound or, conversely, help to break down one compound into many fragments. Living creatures could not carry out normal physiological functions without the intervention of enzymes.

Most proteins are made up from 20 building blocks called amino acids. Amino acids are water-soluble organic compounds made up of the ever-present carbon, hydrogen, nitrogen, and oxygen atoms in varying configurations. (These four elements, incidentally, make up almost 96 percent of the weight of most life forms.) It takes a sequence of three bases on the DNA chain to specify an amino acid. This three-base sequence is sometimes referred to as a "codon."

Proteins may range in size from about 100 amino acids to as many as 1,800. The way they are sequenced, and the three-dimensional shape that they fold themselves into, is critically important to their function. Most bodily processes depend upon the ability of various substances to interlock with each other; whether or not they can do this depends on their shape.

The physical process of producing protein in the cell involves another kind of gene-bearing molecule known as ribonucleic acid, or RNA. (A molecule is a group of chemically bonded atoms.) RNA is very similar to DNA, contains four bases (one of which differs slightly from its counterpart in DNA), does not have a double helix (only a single strand), and is much shorter than a DNA chain. When a protein is to be made, the DNA will synthesize a strand of RNA, and one helix will serve as a template to impress a complementary base sequence on the short RNA strand.

The RNA strand will then carry its message (its protein-specifying sequence of bases) to the production area of the cell (the ribosome), where the protein will be manufactured from the appropriate amino acids. Because of its message-carrying role in protein synthesis, RNA is often referred to as messenger RNA or simply mRNA.

These processes, taking place naturally within living cells, form the foundation upon which the work of genetic engineers is based.

BIOTECHNOLOGY PROCESSES

Microorganisms are responsible for many biotechnological processes. For example, microorganisms are responsible for natural fermentation. The first person to realize that living cells can cause fermentation was French scientist Louis Pasteur, who, in 1857, discovered that microorganisms produce alcohol by the fermentation of sugars. In particular, nutrients on which microorganisms grow are transformed into other chemicals, such as alcohol. The process also often produces a gas; and different phases of the fermentation process result in different end products.

There are other biotechnological processes that include microorganisms. For example, enzymes are proteins used by living cells to break down complex food material into simpler molecules; the converted food releases energy and creates new cell materials. Today, such enzymes are used in cleaning products, including laundry and dishwasher detergents. Biotech enzymes also are used in starch processing, in which huge volumes of corn and other grains are converted to sugars, such as high-fructose corn syrup. And still another example is indigo dye. Denim material uses the largest volume amount of this dye, which is produced by chemical synthesis. (The dye was originally produced from plants.)

Bioremediation, or using naturally occurring bacteria to work in the environment, is also a biotech process. (This is also referred to as environmental biotechnology.) Different organisms need certain types of nutrients, and certain bacteria feed on toxic materials. For example, certain bacteria are used to "eat" hydrocarbons such as oil; one such method has helped to increase the cleanup of spills along coastlines about three to five times faster than conventional means, such as incineration or chemical treatment.

THE GENETIC ENGINEERING PROCESS

While terms like "snipping," "splicing," "chopping," "probing," and "cutting" are used by genetic experimenters, these words usually refer to biochemical processes, rather than to physical actions performed with instruments.

A number of methods of inserting foreign genes into organisms have now been developed, but one of the methods developed in the early stages of research, gene splicing, is still widely used and remains very important. It depends heavily on the use of tiny entities called plasmids.

Plasmids are circular fragments of DNA found outside of the chromosome, particularly in bacteria. They are believed to have specific functions to perform when a bacterial cell meets unusual environmental conditions. In any case, they are able to enter cells, and this ability makes them particularly helpful to experimenters. (Cells are well equipped to keep most foreign agents outside of their protective walls.)

In a typical gene-splicing procedure, plasmids are collected by a centrifuging process (a spinning process that separates material of different densities, e.g., plasma from blood) and brought into contact with a restriction enzyme. The special function of a restriction enzyme is to cut DNA segments at certain fixed points. When the enzyme performs this operation on the plasmids, it leaves them with a gap in their ringed structure.

Foreign segments of DNA, bearing a preselected gene that has been isolated by sophisticated probing techniques from its natural environment, are now introduced into the plasmid population. These newly introduced DNA segments (which might have come from a plant, an animal, or another bacteria culture) now proceed to occupy the gaps in the plasmids and, with the help of another enzyme known as ligase, become firmly attached to them. The newly restructured plasmids (which are known as recombinant DNA) are introduced into a bacteria culture and proceed to enter bacterial cells.

When the cells reproduce (as frequently as every 20 minutes in some cultures), an exact copy of the foreign gene is contained in the new cells.

As the bacteria colony begins to produce protein in accordance with the coding instructions of the foreign gene, it, in effect, becomes a small production "factory" for the desired substance. The "mass-produced" protein can be separated from the cell structure by centrifuging and other chemical techniques.

Many practical engineering problems still remain, centering around impurities, growth rates, and degradation of the product by stubborn cells, but it is reasonable to think that, as with all new processes, these difficulties will be solved in time. This plasmid method is used, incidentally, in the manufacture of insulin for those suffering from diabetes.

Certain viruses (viruses are parasitic microorganisms that are unable to reproduce outside of a host cell) are also used as access vehicles (plasmids and viruses used this way are referred to as "vectors"), because they also have the knack of gaining entry to cells. Some researchers feel that virus vectors offer the most promising path to human cell alteration because of the ease with which they gain entry to cells.

Another method of bringing about changes in the traits of an organism is by directly inducing mutations in its genes. A mutation is an alteration in the gene that changes its function. Mutations are generally harmful. They can result from "accidents" when DNA is replicated or may be caused by radiation or foreign chemicals. If a mutation affects the reproductive cells, it may be passed on to offspring. Genetic engineering, however, has now advanced to the point where it can bring about desirable mutations in specific genes. At present, this effort is largely confined to microorganisms that are important for economic or health reasons. These beneficial mutations may increase the production of some valuable product, for example, or have other desirable effects, such as overcoming or controlling a disease in human beings (sometime in the future). Experimental work has also been done in inducing mutations in insects and mammals. The process of inducing mutations is referred to as "mutagenesis."

One other application of DNA and biotechnology that recently became important is DNA "fingerprinting." The process includes the cross-matching of two standards of DNA. Every living organism (except identical twins) is made up of cells that contain "unique" DNA bases. When the DNA from body tissue (skin, hair) or fluids (blood) are analyzed by certain techniques, a unique genetic pattern is seen. DNA "fingerprinting" is now often used at criminal trials as evidence.

Some people are worried, though, that such biotech techniques may be used against people: could genetic testing eventually be used to exclude people from jobs or health care coverage?

THE FUTURE

Biotechnology has had several major advances, and more will follow. New drugs and processes are being discovered every year.

Things continue to happen fast in the field of genetic engineering, with many discoveries seeming to occur from week to week. The following is only a brief summary of the last decade and a half of genetic research:

- In 1981, a U.S. court ruled that genetically engineered bacteria could be patented.
- In 1982, genetically engineered insulin was approved for use by people suffering from diabetes; a gene for growth hormone was successfully injected into a mouse embryo. The mouse grew to twice normal size.
- In 1984, genetically engineered growth hormone was approved for human use.
- In 1987, gene-altered bacteria was sprayed on a strawberry field in California to prevent frost damage. (This spraying was the first officially approved release of altered bacteria into the environment.)
- In 1989, a foreign gene was injected into a human being (a patient suffering from advanced cancer), with U.S. approval. Also in 1989, the first genetically engineered edible product was approved for human consumption. The product in question, rennin, is an enzyme used in the manufacture of cheese.

- In the same year, the first animal to be covered by a United States patent was a genetically engineered mouse called the oncomouse. It was implanted with a gene that causes cancer, making it useful to those studying the disease in humans.
- On September 14, 1990, a four-year-old child suffering from adenosine deaminase deficiency (ADA) disease became the first human being in history to receive genetically engineered genes. On September 21, 1990, a federal judge ruled that the results of a genetic identification test could be used in his court. This trial marked the first use of such evidence in a federal case. In late 1990, the United States biomedical community commenced work on a 15-year, government-financed ($3 billion) program to map the entire human genome. (A genome is the complete genetic structure of a species.) For human beings, this project involves locating 100,000 genes.
- In March 1991, Michigan State University and the Research Development Corporation of Japan announced a $15-million research program aimed at developing microbe colonies capable of cleaning up oil spills and toxic waste.
- In 1993, researchers "cloned" human embryos. The results triggered a host of ethical questions that continues today.
- In 1994, scientists found a fat-regulating gene shared by mice and people. Researchers also found the first breast cancer gene.
- In 1995, geneticists first sequenced the full genomes of two different bacteria. (Sequencing means mapping the position of genes or chromosomes.)
- In 1996, scientists linked a specific gene to a facet of thought, specifically the ability to visualize and mentally manipulate parts of objects. Other researchers continued to map genomes of various organisms, including a virus that causes skin disease, and baker's yeast, whose cells resemble those of people.
- In 1997, Scottish researchers created a cloned lamb, named Dolly, from the DNA of an adult sheep. In addition, the first human artificial chromosome was created in 1997, and survived for as long as 6 month in cells. Also in 1997, the genome of the "most important bacterium there is," *Escherichia coli*, or *E. coli*, was finally sequenced.

Perhaps sooner than expected, society will be confronted with the difficult questions that rapidly increasing knowledge and advancing techniques will pose in the field of genetic engineering. The questions are not so difficult when we are considering the use of genetically engineered drugs and hormones to treat painful and crippling diseases. The primary concerns in these cases are the safety and well-being of the patient, matters that have always been at the forefront of medical treatment.

Things become more complicated when the course of treatment will include the introduction of foreign or altered genes into the cells of a human being. In the case of patients suffering from a mutation that has caused severe disability, the case for proceeding is certainly strong.

In the case of children suffering from the ADA disorder, which was mentioned earlier, their inherited genes lack the capacity to produce a particular enzyme (adenosine deaminase), which is vital to the functioning of the immune system. The slightest infection can be life-threatening to them. This affliction is similar to the disease that caused a young Texas boy to spend his entire brief life in a plastic bubble chamber. Those who suffer from ADA disorder are currently treated with periodic injections of a synthetic version of the missing enzyme, but they still live in constant fear of infection. Nevertheless, there is great concern about the risks and consequences of treating them by inserting foreign genes into their cells.

Further down the road, and even more difficult and soul searching, is something that has already occurred with the cloning of the sheep in Scotland. If humans were cloned, what rights would the clones have? Would they be considered expendable and used for hazardous work? Who would be cloned? What laws will we have to pass to control cloning research, especially if humans are involved? Besides the cloning question, genetic engineering brings up a number of other sensitive issues. What will be the potential of genetic engineering for "improving" human beings, i.e., making them more "intelligent" or "stronger" or "healthier"? Is this something we should even consider? These possibilities raise profound philosophical, religious, social, and ethical issues that touch the sensitivities of everyone. At some point they will have to be addressed.

FURTHER READING

Books

Altered Fates: Gene Therapy and the Retooling of Human Life. Jeff Lyon, Peter Gorner. New York: W.W. Norton & Co., 1995.
> A well-written volume that deals with the history and possible future of gene therapy.

Biotechnology. John E. Smith. Cambridge, England: Cambridge University Press, 1996.
> A detailed account of the applications in biotechnology.

Biotechnology from A to Z. William Bains. Oxford, England: Oxford University Press, 1993.
> An introductory guide to ideas and terminology in biotechnology.

Introduction to Genetic Engineering. William H. Sofer. Newton, MA: Butterworth-Heinemann, 1991.
> An introduction to the foundations of molecular biology and molecular cloning.

The Science of Jurassic Park and the Lost World: How to Build a Dinosaur. Rob Desalle, David Lindley. New York: Basic Books, 1997.
> A popular look at the plausibility of the Jurassic Park premise—building a dinosaur from DNA preserved in amber.

Understanding DNA and Gene Cloning: A Guide for the Curious. Karl A. Drlica. New York: John Wiley & Sons, 1996 (3rd edition).
> A look at molecular biology and recombinant DNA technology for the layperson.

Articles

"Dr. Tinkertoy." R. Pool. *Discover*, February 1997, 50-54.
> N. Seeman's work in genetic engineering.

"Finding a Cure in DNA?" N. Freundlich. *Business Week*, March 10, 1997, 90.
> Genetic engineering and medicine.

"The Ultimate Medicine." Geoffrey Montgomery. *Discover*, 1990, 60-68.
> Treating genetic diseases in human beings through the new techniques.

Chip

BASIC DESCRIPTION

A chip is a tiny (typically about a third of an inch square or smaller), regularly shaped strip of silicon (or similar material), densely packed with miniature electronic components. These components are integrated with each other to perform control, logic, and/or memory functions. The complete chip, covered by a protective plastic package and sporting many spider-like gold-wire connectors, can be thought of as an electric circuit with switches, cables, and relays, but on a tiny, almost microscopic scale. They use very little power and perform their functions with great speed. Chips can be wired to host devices such as computers, automotive systems, telephones and telephone switchboards, and many household appliances, which they will then monitor or control.

Note: Chips are also referred to as integrated circuits or microchips. Microprocessors are the high-capability chips that people are referring to when they speak about a "computer on a chip."

REPRESENTATIVE APPLICATIONS

Automobiles

Dedicated chips, i.e., chips designed to perform certain specific functions such as optimizing the fuel mixture, are already commonplace in automobiles.

Cameras, Calculators, and Watches

As in cars, dedicated chips, drawing their power from batteries, operate key functions of these everyday devices.

Computers

Chips are the essential building blocks of most computers, performing or controlling virtually all of their functions.

Industry

It is fair to say that any recently installed industrial equipment of any complexity depends on chips for some of its functions.

Medicine

Pacemakers, hearing aids, and virtually every other medical device of any complexity now contain one or more chips.

Space Program

Most of the spacecraft in the space program—from the space shuttles to the interplanetary probes—rely heavily on chip technology.

General

Many household appliances, such as refrigerators, washer-dryers, microwave ovens, TV sets, VCRs, stereos, alarm systems, facsimile (fax) machines, and telephones (including portable and cellular phones), now have chips built into them. Such chips permit the appliances to automatically perform functions that previously had to be done manually by the owners. Chips also enable some of these appliances to perform brand new functions (such as timing the length of a telephone call or determining the caller's phone number) which increase their usefulness to their owners. It is reasonable to say that there is no significant commercial product that has not been affected to some extent by the "chip revolution."

BACKGROUND

When the transistor (+) was invented in 1948, it transformed the computer from a huge, power-hungry, unreliable, heat-generating apparatus into a smaller, faster, more dependable, cooler-running device. On a smaller scale it did the same thing for radios, and, for the first time, the portable radio became a truly practical device.

By replacing the fragile, high-temperature vacuum tube in most applications, the transistor opened the door to countless new products. But the onrush of technology that followed the development of this new device was hampered by bottlenecks in the manufacturing process. As the components going into various products became smaller, the problems associated with wiring them together became a significant factor. The computers of that period required more than 20,000 transistors, 100,000 diodes (+), and thousands of other components, all of which had to be wired into the circuits. Doing this assembly by hand was extremely expensive and resulted in a high rejection rate.

Industry was well aware of these problems and a number of solutions were proposed, none of which achieved any great success. Finally, two individuals, one at Texas Instruments and the other at Fairchild Semiconductor, working independently, developed a concept that eventually proved to be the answer to this manufacturing problem.

In September 1958, Jack Kilby of Texas Instruments (TI), in a meeting with his management, demonstrated a working model of a circuit built on a small piece of silicon. Some time later, Bob Noyce, one of the founders of Fairchild Semiconductor Corporation, working without any knowledge of the developments at TI, developed a similar device. A period of litigation over patent rights ensued, but the dispute was eventually settled amicably, and both men are now regarded as co-inventors of one of the great technological concepts of the century.

THE SIGNIFICANCE OF THE CHIP

The development of the chip has made it possible to build more and more processing power (and memory storage) into smaller and smaller packages. Modest-looking desktop computers now have enormous capability due to their efficient processing and memory chips. Tiny hand-held calculators depend on more modest chips, but many of them provide the kind of power that was only available from desktop computers just a few years ago.

How the Chip Works

Circuits on chips comprise the basic operating structure of most computers and other logic-based devices. The operation of these circuits rests on a set of logic steps that are simple in principle but complex in practice. Arrays of switches (called gates) are placed strategically throughout the circuitry to allow, or prevent, the passage of electric pulses. The pulses passing through the system represent bits of data; joined to-gether in appropriate patterns, they embody the information that the computer is processing at any particular time.

Some gates will allow an output pulse each time they receive an input pulse; others may require two (or more) inputs before allowing an output. Other such combinations of pulse transfers cause predetermined effects. Taken together, these millions of split-second gate events enable the computer to perform its logic operations.

In noncomputer applications, the availability of mass-produced, general purpose chips, built to published specifications and sold "off the shelf," makes it possible for manufacturers to design them into a wide range of products whose usefulness and versatility can thereby be enhanced significantly. Some of these chips may do arithmetic functions and go into hand-held calculators, others may perform timing functions or similar operations. The manufacturer who buys them will integrate them into a marketable product.

Creating a Chip

Considering that the average chip is smaller than a thumbnail and may contain millions of components (interconnected transistors, diodes, and resistors), the fact that it can be successfully manufactured at all is a tribute to the ingenuity of the engineers and scientists who worked out the process.

So difficult was the task, in fact, that it took almost two years after the announcement of the new device in 1959 before the first chips appeared on the market. (IBM did not use chips in its mainframes until 1969.)

The Manufacturing Process

Design—A chip must begin with a design. An engineer or researcher determines what the chip has to do and then prepares a schematic drawing containing the main features of the circuit. The actual design of the circuit begins at the computer with the help of a CAD (Computer-Aided Design) (+) program.

The use of a CAD program is an example of an expert system application and is one that has become indispensable to the integrated circuit design process.

The CAD program analyzes the designer's concepts and requirements; performs essential calculations; and allows for material limitations, operating constraints, and manufacturing and process characteristics. In general terms, it takes all this information and uses it to select the optimum paths toward the designer's goals. The whole process is a kind of dialogue carried on between the designer and the computer program through the keyboard, the monitor screen, and the printer. It is full of "what if" questions and "if this, then not that" answers.

After the design has been completed and tested for functional consistency by a computer simulation program, a computer-driven plotter makes a set of large-scale layout drawings of the circuit.

Since the layout drawing is (at least) several hundred times larger than the actual circuit, it has to be reduced to near-actual size before it is used in the masking process (see below). A very large, vibration-free projection camera of the type used in the printing industry is used to reduce the size of the drawings.

Masking—The actual chip circuit is formed by carefully positioning multiple layers of material, one upon another, onto the silicon wafer to form a three-dimensional circuit. The silicon wafers that will carry the components are produced in a "clean room," an area with extremely low humidity and in which each cubic foot of air must contain fewer than 1,000 tiny specks of dust. This cleanliness is necessary to ensure that no dust or particles end up on the silicon surface during the processing of the chip.

The transfer of the circuit design elements to the multiple layers is achieved with the use of extremely thin sheets of glass referred to as "masks." The images of the circuit, in their final actual miniature size, are first transferred photographically to the masks, which will have been coated with a light-sensitive emulsion. Because the silicon structure onto which the circuit patterns will be transferred is a flat circular wafer large enough to be cut up into hundreds of individual chips, the masks must contain not just one image of the circuit, but hundreds, in order that a single exposure of the mask will transfer all the images at once. (See figure 1.)

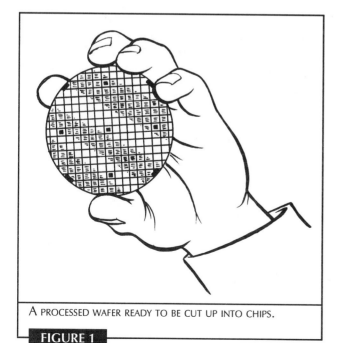

A PROCESSED WAFER READY TO BE CUT UP INTO CHIPS.

FIGURE 1

This step is accomplished by a process well known in the printing industry. Each image is placed on the individual mask by a step-and-repeat procedure that moves the mask around under the projection camera, stopping momentarily for an exposure at each position before moving to the next position.

At appropriate stages in the manufacturing process then, the proper mask is precisely positioned over the silicon wafer, and intense light from a projector is directed through the mask to imprint the hundreds of circuit outlines onto a light-sensitive coating (photoresist) on the surface of the wafer. In printing, this process is known as contact printing. In chip processing, projection lithography and electron beam lithography are alternate methods of achieving the same result.

Processing—The exposed photoresist on the surface of the wafer is then developed and the surface etched to reproduce the pattern of the circuits.

"Doping"—Carefully controlled, minute quantities of impurities such as phosphorous, boron, or arsenic are diffused through targeted sections of the surface pattern in order to cause selected segments of the wafer to become conductive. This conductivity is achieved because the atoms of the diffused impurities combine chemically with the silicon atoms in such a way as to leave free negative or positive charges in the treated segments of the wafer. These free charges will carry electric current if a voltage is now applied to the circuit. Within these "doped" areas, tiny transistors (+) and other components are formed within the silicon itself.

A metal coating (usually aluminum slightly alloyed with silicon and/or copper) is now deposited on the surface and etched to form the connections between the circuit devices. Some metal may also be used to form components or parts of components.

Many passes through this basic cycle are needed to get all elements of the circuits into place. After all the processing has been completed, the wafer is tested to determine if any defective circuits are present. It is then cut up into individual chips. Defective chips are discarded, and the rest are protectively packaged.

The entire manufacturing operation must be conducted under conditions of extreme cleanliness. Even tiny dust particles can cause chips to malfunction. Workers wear head (and sometimes face) coverings and special lint-free uniforms. Machines are carefully protected to prevent them from releasing contaminants into the work area. The air in the workroom is filtered and regularly monitored for contaminants.

In addition to cleanliness, the intricate machinery used in the manufacturing process must be completely free from vibrations and other disturbances. Although they can be mounted on special shock absorbers, these mechanisms are so sensitive that certain locations, because of nearby traffic or other sources of vibration, simply cannot be used.

The process entails large capital outlays and high worker skills; however, when the equipment setup and worker skills are achieved, chips can be successfully mass-produced at low unit cost. One great advantage of this manufacturing process is that a very high degree of uniformity among transistors can be achieved because all components are formed within the same block of silicon.

THE FUTURE

There has been a saying in the industry that the number of components on a chip doubles every year without an increase in the size of the chip. This adage was actually true for most of the 1970s but is now a more difficult objective. It does, however, express the goal that is pursued with relentless determination in the industry: ever more processing ability in smaller and smaller physical space. The problem that researchers now see on the horizon is how to conquer the physical limitations of packing more circuits onto smaller chips, such as the effects of interference, heat, or current flux between the smaller components.

As integrated circuits grow ever tinier, the need for sharper and better defined image patterns grows more urgent. High frequency electron beams and "soft" X-rays are now being used to attain greater resolution in creating circuit images. With better techniques being developed, such as the use of the scanning tunneling microscope that moves individual molecules on a silicon surface, smaller circuits may be attainable. Another method of moving atoms is with electron-beam lithography. Recently, scientists at the Cornell Nanofabrication Facility created the world's smallest silicon device with this method—a 10-micron-long "nanoguitar" that actually has strings that can be plucked with an atomic force microscope (but the sound is inaudible). This experiment is another step toward smaller chips and components.

Materials other than silicon are also getting increasing scrutiny as basic chip material. Gallium arsenide, although it has been used on a small scale for many years, never attained the acceptance of silicon because it is harder to handle in the manufacturing process. However, it is being looked at more closely these days because more components can be etched onto its surface, it can run at lower temperatures, and it has a faster switching speed than silicon. It is also

based on quantum wells, in which the 1s and 0s (see binary numbers [+]) are represented by the flow of singular electrons (which will create a "cleaner" signal), not the bulky flow of multiple electrons as with silicon.

Traditionally, chips have been placed in four rough size categories: SSI for small scale integration, MSI for medium scale integration, LSI for large scale integration, and VLSI for very large scale integration. In recent years the acronym ULSI for ultra large scale integration has come into use. The boundaries between these categories have been rather loosely defined, but ULSI units are now packing millions of components on a single chip. The newest acronym is the GSI chip. GSI stands for gigascale integration (giga is the metric [+] term for billion) and represents a long-term target for the industry. Radical innovations in the masking and processing methods will have to be developed before the resolution needed for circuits of this size can be attained. However, some experts feel that the billion-component chip will be achieved by the beginning of the twenty-first century.

Even further down the road, integrated optical chips may become a realistic possibility. These chips would hold the optical equivalents of transistors and other components. Since their medium would be light, instead of electrical current, they would generate little or no heat; they would provide faster switching; and they would be immune to electrical noise. Although many significant engineering problems remain to be solved, some experts feel that light-driven chips will represent the next radical advance in computer technology.

Another technology being explored is biochips, in which organic (carbon-based) materials replace the memory chips inside the computer. Several biochip studies have been conducted; for example, a protein derived from a type of saltwater bacteria that responds to light changes was used to simulate an on-off state in a computer.

FURTHER READING

Books

Microchip Fabrication: A Practical Guide to Semiconductor Processing. Peter Van Zant. New York: Semiconductor Services, 1996.
   An introduction to the entire chip production process written in nontechnical language.
The Microprocessor: A Biography: Silicon Valley Series. Michael S. Malone. New York: Springer-Verlag, 1995.
   An understandable explanation of how microprocessors are made and work.

Articles

"Birth of a Chip." L. Gwennapil. *Byte*, December 1996, 77-78.

How chips are made today.

"How Microchips Shook the World." P. Wayneril. *Byte*, December 1996, 68-70.

More on the chip revolution.

"Microscopic Mass Production." F. Saunders. *Discover*, February 1997, 30-31.

An update on how polymer molds are used for making computer chips.

Cryogenics

BASIC DESCRIPTION

Cryogenics is the process of reducing the temperatures of selected substances to extremely low levels (near absolute zero in some instances) in order to utilize the resultant changes in their physical properties to achieve significant engineering, industrial, medical, and research objectives. Extremely low temperatures are achieved by liquefying gases. The field of medicine is looking to cryogenics for better and completely new methods of preserving major human organs for transplant, and it also uses cryogenic cooling in some diagnostic techniques such as MRI (+). Energy specialists hope that cryogenic cooling will eventually lead to the more efficient use of electricity as well as better methods of storing energy. Cryogenic temperatures are generally considered to be those in the range below minus 100 degrees Celsius (minus 148 degrees Fahrenheit).

REPRESENTATIVE APPLICATIONS

Agriculture

Long-term storage of the sperm of large farm animals by cryogenic methods has substantially altered breeding practices and economics in cattle and dairy farming. It is estimated that three quarters of all cattle in Great Britain are the result of artificial insemination using frozen sperm.

Biology

The ability to freeze and store microorganisms has enabled researchers to preserve pure strains of particularly important types of bacteria. The ability to store animal sperm has enabled scientists to interbreed creatures whose natural mating cycles occur at different seasons of the year.

Commerce-Industry

The liquefaction of gases by cryogenic techniques has vastly simplified procedures for the storage, shipment, and distribution of large quantities of commercially used gases. The expanded use of oxygen in the steel industry has made it necessary for this gas to be available for shipment and use in liquid form. The liquefaction of natural gas has made it practical to utilize this important resource far from its point of origin.

Medicine

(a) Cryosurgery uses extreme cold to seal off blood vessels, creating, in effect, "bloodless surgery."

(b) The long-term storage of human blood cells, corneas, bone marrow, and sperm has been achieved through cryogenics. Critically needed are cryogenic procedures for achieving long-term storage of major human organs. This procedure will make it possible to obtain much more precise matches between donors and patients, greatly enhancing the success rate of implant procedures. The cryostorage of major organs has proved to be a difficult problem, however, because the formation of ice crystals tear apart the fragile cells in tissues, and because not all parts of the organ freeze at the same rates. Progress to date to freeze organs has been somewhat disappointing.

Space

Gases such as oxygen and hydrogen, liquefied through cryogenics, are used as rocket fuels. The use of liquid hydrogen fuel was a factor in the success of the Apollo missions to the moon. Cryogenics is also important to planetary spacecraft (+) that need cooled components to counter the effects of thermal cycling by the heat of the Sun and coldness of space.

Superconductivity (+)

The discovery of superconductivity, i.e., the ability of certain supercooled materials to conduct electric current without resistance, has become a very important topic in the high-tech world. An entire section has been devoted to it later in this book.

BACKGROUND

More than two hundred years ago, the great French chemist, Antoine Laurent Lavoisier (1743-1794), wrote:

> If the Earth were taken into a hotter part of the solar system . . . all our liquids and even some metals . . . would become gases . . . . If on the other hand . . . we were taken into...colder regions . . . for instance, near Jupiter . . . our rivers and oceans would become solids . . . and our atmosphere would become a liquid . . . .

We tend to think of materials as either solids, liquids, or gases when they exist in such states at everyday conditions of temperature and pressure. In fact, however, most materials will convert from their "everyday" state to one of the others if appropriate changes are made in their temperature and pressure. For example, helium gas will become a liquid at minus 269 degrees Celsius (minus 452.2 Farenheit); water is a solid below zero and a gas above 100 degrees Celsius (212 Farenheit).

Temperature is a measure of the average motion of a substance's molecules; if they are highly active, the substance (relatively speaking) is hot; if they're inactive, it's cool. Cryogenic researchers tend to use the Kelvin scale to measure temperature. The Kelvin scale is the official metric (SI) (+) scale for measuring temperature. It starts at absolute zero, which can be informally defined as the temperature at which all physical activity ceases. The Kelvin scale moves upward from absolute zero in increments equal to Celsius degrees, so that at 273 kelvins (or 273.16, to be exact) the freezing point of water is reached. In other words, 273 kelvins equals zero degrees Celsius or 32 degrees Fahrenheit.

Lavoisier's intriguing speculation about rivers becoming solids and air becoming a liquid was an inspiration for those relatively few nineteenth-century physicists and chemists who had dedicated themselves to investigating the behavior of materials at very low temperatures.

In 1877, Louis Paul Cailletet, a mining engineer from Chatillon-sur-Seine, announced at a meeting of the Academy of Sciences in Paris that he had liquefied oxygen. Within a few weeks, he followed up by announcing that he had liquefied nitrogen. Since ordinary air is composed of approximately 80 percent nitrogen and 20 percent oxygen, he had effectively liquefied our atmosphere, thus fulfilling one important part of Lavoisier's prediction.

Cailletet's process, however, produced only faint droplets of these liquefied gases. In working with oxygen, Cailletet's method was to accumulate it in a very sturdy vessel and cool it by evaporating sulfur dioxide in contact with the vessel. He then compressed the oxygen as much as his equipment allowed. At this point he released the pressure suddenly, allowing the oxygen to expand, and this sudden cooling produced a few droplets of liquid oxygen.

It was not until 1883 that significant quantities of liquid oxygen and nitrogen were produced. This feat was accomplished by two Polish researchers in Cracow. Szygmunt von Wroblewski and K. Olszewski succeeded in this effort by making improvements in Cailletet's equipment. In 1898, James Dewar, a distinguished British scientist, announced the liquefaction of hydrogen.

Research efforts then focused on helium, which is a fairly rare element, and one that was known to have an extremely low liquefaction temperature. In July 1908, Professor Heike Kamerlingh Onnes of the University of Leiden in the Netherlands announced that he had liquefied helium at 4.2 kelvins on the absolute scale. This reading was the lowest temperature ever achieved in any laboratory to that date and represented a significant milestone in the history of cryogenics. Dr. Kamerlingh Onnes, who is considered by many to be the greatest of cryogenics researchers, later went on to discover superconductivity by cooling mercury with liquid helium. He received the Nobel Prize in physics in 1913.

THE COOLING PROCESS

An important method of achieving cryogenic supercooling is by liquefying gases and then using the liquefied gases to cool other substances.

One system used in industry is called "cascading." First a gas that can be liquefied by pressure alone is converted into liquid form. Then a vessel containing a second gas, one that requires a lower temperature to become a liquid, is immersed in it. The already liquefied gas cools the other gas and causes it to liquefy. This process is then repeated in successive steps until the targeted gas is liquefied.

Judicious selection of the interacting gases is necessary to assure the success of the process. When the targeted gas is liquefied, it can be cooled further by pumping off the vapor that evaporates from its surface. This decreases the pressure in the system and results in additional cooling of the liquid.

Cooling is achieved in the ordinary household refrigerator by using a motor-piston combination to compress the gas and then releasing the heat extracted from the compressed gas into the kitchen. The compressed gas is then allowed to expand, thereby reducing its temperature and permitting it to cool the contents of the refrigerator. It is then compressed again, and again permitted to expand. This repetitive cycle results in a stable, cool temperature inside the refrigerator.

Maintaining a stable supply of a very-low-temperature liquefied gas is normally accomplished by storing it in a dewar. A dewar (named after the liquefier of hydrogen) is a double-jacketed container with a vacuum between the inner and outer jackets to prevent heat passage. Helium, because of its extremely low critical temperature (the lowest of any known gas), is normally transported in a dewar surrounded by a second dewar containing liquid nitrogen.

Since the 1930s, cryogenic cooling below 1 kelvin has depended on magnetic techniques. In 1926, William Francis Giauque, a young Canadian chemist, reasoned that if the disorderly spin of electrons (+) in a substance could be "calmed down," a significant temperature reduction could be achieved.

He selected gadolinium sulfate for his experiment because of its favorable magnetic properties. A sample of this salt was reduced to a temperature of 1 kelvin through liquid helium cooling and was then subjected to a magnetic field. This caused the randomly spinning particles to line up in orientation with the magnetic field. This step has been likened to compressing a gas. The thermal energy that had been concentrated by this process in Giauque's experiment was then carried off by the liquid helium. At this point, the magnetic field was cut off, and the tendency of the particles to resume their random spin was offset by a sharp drop in temperature. This step is compared to allowing a compressed gas to expand.

Magnetic cooling such as that described above has led to temperatures within a tiny fraction of a kelvin of absolute zero.

THE FUTURE

In the field of medical biology, the search for methods of storing major human organs for extended periods will continue and intensify. The major problem to be overcome is the formation of tissue-damaging ice crystals when the organs are cooled to cryogenic temperatures. Certain cryobiological studies are being conducted to solve this problem, including research into insects that use certain natural cryosubstances that allow them to overwinter in cooler climates.

For space purposes, scientists hope to use cryobiological techniques to store astronauts' blood for emergencies in space. In addition, certain cryostorage techniques for foodstuff are being improved for the space program. In particular, experiments on the Long Duration Exposure Facility, which orbited the Earth for many years, proved that certain seeds could be stored in the cold of space without affecting their growth capabilities.

The continual threats of disruption in the oil supply have led to increased interest in alternative fuel sources. One promising possibility is liquid hydrogen, which has been successfully used as a fuel in space vehicles and may have far wider application, including possible use in automobiles.

Both "high" and "low" temperature superconductivity (+) promise to play important new roles in industry and science.

Efforts to bring substances closer and closer to the goal of absolute zero will continue. In 1995, several researchers at the National Institute of Standards and Technology at the University of Colorado managed to cool rubidium atoms in a new type of "atom trap," recording a temperature of 200 nanokelvins, close to absolute zero. The atom trap allowed the fastest (or hottest) atoms to escape, leaving behind the slower (cooler) atoms. The researchers then measured the speeds of the atoms using lasers (+) to make the atoms fluoresce (emit light), then videotaped the motions, which allowed them to infer the temperature. Researchers hope that such cold atomic clouds can be used to increase the precision of atomic clocks and improve various types of spectroscopy, as the slower moving atoms may be easier to "control."

In basic cryogenics research, much future effort will center around the continued study of the strange properties and behavior of materials at super low temperatures.

FURTHER READING

Books

Matter and Methods at Low Temperature. Frank Pobell, Seth Luth. New York: Springer-Verlag, 1996.
  A technical look at low-temperature physics.
The Quest for Absolute Zero: The Meaning of Low Temperature Physics. K. Mendelssohn. New York: McGraw-Hill, 1966.
  An authoritative discussion of the history and theory of cryogenics.

Articles

"Cold Storage." Philip E. Ross. *Scientific American*, February 1990, 20-22.

A brief news article updating progress and problems in the preservation of human organs.

"Freeze-Drying." H. Aschkenasy. *Scientific American*, September 1996, 184.

Recent techniques in cryoresearch.

"Frozen Future." A. Stuttaford. *National Review*, September 2, 1996, 104.

Freezing in the future.

"30 Years of Cryogenics." J. Gardner. *Cryogenics*, January 1990, 3-4.

A short but informative editorial reviewing developments in cryogenics over the past 30 years.

Digital Image Processing

BASIC DESCRIPTION

Digital image processing (DIP) is the technique of converting a photograph or other image into digital form by assigning numbers to its visual characteristics and then processing those numbers in a computer. The conversion of the image to numbers is referred to as "digitizing" it. This process involves the use of a scanner that directs a spot of light over the entire image in a sweeping grid pattern, monitors the reflections from the image, and assigns numbers to the reflections according to their density and color. Subsequent computer processing may include enhancing the original picture (a sort of electronic touch-up) and correcting structural errors arising in the image-acquisition process. All of these processes take place within the computer, which uses algorithms (sub-programs that perform specified operations on the digitized picture elements) to achieve the desired results.

REPRESENTATIVE APPLICATIONS

Art Conservation

Some museums have begun to digitize art masterpieces to clarify (through computer analysis) fading details and to assure that these great paintings will survive in some form, even if the originals are eventually lost to the ravages of time.

Commerce

Ordinary snapshots may be enhanced through digital processing techniques. Because of current high cost, this practice is not yet widespread, but prices for the images and equipment are continually dropping. Digital processing is already playing a role in commercial photo finishing.

Finance

Many banks and other financial institutions are now digitizing checks and transaction receipts and are sending computer-generated copies of these documents to their clients instead of originals. This process saves a great deal of handling and mailing costs and has proven popular with clients.

Mapping

Digitized maps can be reoriented, adjusted to accept information from other sources, and enhanced. Digitized maps are also used for global positioning systems (+), in which a person's location can be easily determined by the press of a button.

Media

Old black-and-white movie films are being "colorized" through digital image processing techniques. Such techniques are also being used to enhance movie graphics and special effects.

Medicine

Medical imaging (+) techniques make increasing use of digital image processing. Among the earliest applications was the analysis of X-ray images. Currently, CT (+) scans and the growing area of specialized imaging techniques, such as MRI (+) and ultrasound (+), are dependent on digital image processing.

Meteorology

Weather satellites are constantly making "readings" of weather patterns in a variety of ways. These patterns can be digitized for analysis, transmission, and enhancement.

Military

The digitizing of aerial reconnaissance photographs for transmission and analysis is routine.

Research

Many patterns obtained from instrumentation readings are digitized and then analyzed by digital image processing techniques in science laboratories throughout the world.

Space Exploration

Space exploration devices were among the first to use digital image processing. These techniques are now indispensable for obtaining imagery data from space.

BACKGROUND

The Voyager 1 and 2 spacecraft both started their journeys to the outer solar system in 1977, with Voyager 1 flying past Jupiter and Saturn, and Voyager 2 flying past Jupiter, Saturn, Uranus, and finally, in 1989, Neptune. Their assignments were to shoot close-up pictures (close-up by space standards) of the planets and their satellites and send them back to Earth.

Clearly, the task of taking the pictures would be difficult enough by itself. The Voyagers each carried two TV cameras, one wide angle and one narrow angle; these cameras would have to be aimed and carefully exposed, with proper allowances made for the movements of the spacecraft and the planet. Once taken, the images would have to be transmitted by radio, across the vast, empty, multi-billion-mile expanse of space that separated the Voyagers from the surface of the Earth.

The first pictures taken in space (by the Soviet and early U.S. probes) were analog (+) images. While it is possible to enhance analog images and correct distortions in them, the process is limited and difficult. And while it is possible to transmit weak analog signals through space for relatively short distances, such as that from the Moon to the Earth (about 370,070 kilometers [230,000 miles]), the greater distances to the planets (an average of 77,232,000 kilometers [48,000,000 miles] to Mars, for example) are such that analog transmissions would fade and be lost, merged into the background noise of space.

Scientists at NASA's Jet Propulsion Laboratory in Pasadena, California realized that they would have to turn to digital techniques to carry out further exploration into deep space. They accomplished this objective with the Mariner 4 probe, which visited Mars in 1965 and sent the first digitally transmitted images back to Earth. The results were highly successful and set the pattern for future space journeys.

The spectacular images transmitted by Voyager 1 and 2 to Earth from the many planets and satellites of the outer solar system are among the crowning achievements of digital image transmission. (See figure 2.) And since that time, numerous planetary spacecraft—from the Magellan at Venus to the Pathfinder on

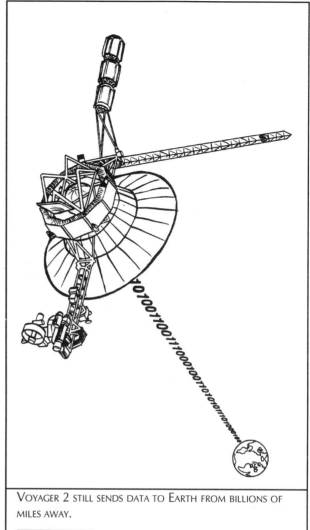

VOYAGER 2 STILL SENDS DATA TO EARTH FROM BILLIONS OF MILES AWAY.

FIGURE 2

Mars—have continued the tradition of sending back amazingly detailed digital images from the other members of our solar system.

THE PROCESS

The digitizing process must begin with an image or a pattern. The image may be obtained with a conventional camera using ordinary film or with a TV camera, an X-ray machine, or some other sensing device. Or the image may simply be a light pattern reflected from an object like a planet. In the broadest sense of the word, the image may be nothing more than a pattern of signals, meaningless or invisible to human eyes until processed by a computer and converted into visual form.

One of the most familiar everyday images is a simple snapshot. To digitize a snapshot, we could conceivably divide it up into a tiny checkerboard pattern and then assign a number to each square. For example, the darkest squares would get a zero, the lighter squares, some positive number proportional to their brightness. Following this procedure, we would end up with a matrix of numbers. (A matrix is simply an orderly grouping of numbers linked to each other by some relationship.)

If we were to hand this matrix to someone, along with a scale, i.e., a table, matching numbers to shades of lightness and darkness, it should be possible for that person, working with a drawing pencil, to come up with a fairly good reproduction of the original picture. We can assert with some confidence that this procedure would work, since the whole process is similar to the number-coded painting sets that were popular years ago.

With the help of some very sophisticated hardware and software, and allowing for some formidable engineering complications, images are digitized, in principle, in accordance with the simple procedure described above.

If we run the original snapshot through a modern digitizer, our manual "checkerboard" technique is replaced by some form of scanning device. The scanner generates a tiny beam of light and moves it in continuous sweeps in a grid pattern across the surface of the photograph. The scanner has a monitor that senses the differing values of the reflected light as the beam contacts white, black, and in-between spots on the surface. These values have numbers assigned according to their visual characteristics and are stored for processing. All of this scanning takes place at high speed.

It can be seen from this description that digitizing is a sampling process, since it is impossible to make the size of the "squares" small enough to cover every variation on the surface of the photo. The tiny sampled areas are called "pixels" (for picture elements), and their size determines the "resolution" (or sharpness) of the digitized image. The smaller the size of each pixel or, to say it another way, the greater the number of pixels per area of image, the greater the resolution (or sharpness) of the reproduced image.

If the image to be digitized is in color, additional matrixes have to be generated to assign numbers to the color values.

Once an image has been adequately digitized (clearly, "adequate" will vary with the application: medical and scientific work will require very high resolution; snapshot enhancement much less), the pixel numbers can be entered into the computer for processing.

Processing can be divided into two broad categories:

1. Those modifications made in an image to emphasize its good points and lessen its bad points. Broadly speaking, this type of processing is known as "enhancement." The important point is that there is no intent to change the character of the picture, only to make it more pleasing or more informative. Examples include improving contrast or eliminating fuzziness in a snapshot. For instance, if the fuzziness is due to noise, i.e., tiny unwanted spots, the spots can be removed, or their impact reduced, by putting the values of groups of pixels in order and using the median value for all pixels in the group. This process is called "noise smoothing."

2. Changes made to correct what could be called structural problems in the image acquisition process. This type of processing is sometimes referred to as "restoration." All types of cameras, monitors, and sensors (+) have their peculiarities and weaknesses. In addition, the conditions under which image acquisition occurs may cause geometric distortions or similar difficulties in the final result. When these conditions are understood, they can be compensated for by using mathematical techniques in processing, e.g., eliminating the barrelling effect (straight lines on the borders of the image appear curved) that occurs with some electronic cameras.

In both (1) and (2), the changes made are generally extrapolations from the information in the image. (Extrapolations are, in a sense, "educated guesses," in the form of projections from what is known for sure.) In the case of (2), new information is sometimes introduced into the image, as, for example, in the form of a template, highlighting certain features, overlaid on the original picture.

In addition to the above procedures, which, generally speaking, are efforts to make the processed image as accurate a depiction of the original object as possible, it is possible to go far beyond mere accuracy. Deliberate distortions can be introduced, or images from other sources can be combined with the original to create all kinds of special effects.

Old black-and-white movie films have been "colorized" so that they look quite natural to people who are seeing them for the first time. This process involves newly developed techniques that rely heavily on digitizing methods.

Digital processing techniques can also make faked photos almost impossible to detect. In the jargon of the trade, they are "seamless."

Interestingly, scientists sometimes introduce false coloring into images transmitted from space to facilitate the analysis of surface features.

In computer-aided design (CAD) (+) programs, the computer can produce images based entirely on numbers entered by the human designer.

Since digitized images consume huge amounts of space in computer memory, techniques have been developed for compressing the data needed for image storage. These techniques take advantage of the fact that most pictures contain areas where there is virtually no contrast. For example, if a picture has a large area of solid black, there is no real value in storing individual numbers for each pixel in that area (obviously such numbers would all be the same). In such a case, the quantity of stored numbers can be reduced by a technique of selective geometric distortion. This technique shrinks the low contrast areas and leaves the other areas alone. When the image is reproduced, the process is reversed and the shrunken areas are returned to their original shape.

In summary, our ability to transform pictures (and even nonvisual signals) into numbers that can be processed in a computer opens the door to many interesting possibilities.

Once in the computer, sub-programs, called algorithms, can operate on the numbers to (a) improve the appearance or accuracy of the original image, (b) transform it (for scientific or aesthetic reasons) into something bearing little or no resemblance to the original image, or (c) create an image where only a nonvisible pattern of signals existed before.

The processed image may be displayed in a variety of ways; for example, it may be displayed on a computer or TV screen, it may be printed as a hard copy such as a photograph, or it may be projected onto a movie screen.

THE FUTURE

Use of digital image processing continues to grow rapidly in industry, science, space, and financial applications.

In the early 1990s, the Eastman Kodak Company announced the introduction of a CD snapshot storage system that combines silver halide photography with DIP techniques. In this new system, people take pictures with conventional cameras, but, at the photofinishing plant, the film is digitized. Prints are made from the digitized image, and that digitized image is also be stored on a compact disc. With special equipment it is possible to view the pictures on a regular TV set. In addition, copies of the original snapshots can be made directly from the compact disc by the photofinisher.

Today, the concept of digitized images has grown exponentially, especially since commercial digital cameras have become more available to the public. Similar to the CD storage system mentioned above, these digital cameras make it possible to view the images on a computer screen, allowing the client to choose the desired photos without processing the entire roll of film first.

However, it is in the area of health care that some of the most important applications of digital image processing have taken place, and it is within this area that dramatic future developments may be anticipated. In addition to the problems of creating and processing images, health-care facilities face enormous problems in storing and transmitting the images after they have been created.

PACS (Picture-Archiving and Communications Systems) have recently been introduced in the medical centers of about a dozen U.S. universities. This system will digitize and then store images, pictures, and text associated with medical care. It will also provide networking so that images and other data can be forwarded to any point within the system. It will mean enormous savings in storage space and much quicker access to needed data. The PACS system will draw heavily on DIP and networking (+) techniques.

Such storage techniques have also made an impact on the Internet, in which multitudes of digital images, including those digital images taken by planetary spacecraft, are made available to the public for viewing or downloading. Other helpful images include those available from the Visible Human Project by the National Library of Medicine, in which detailed images of cross-sections of a male and female are available for anatomical study.

FURTHER READING

Books

Digital Image Processing. Kenneth R. Castleman. New York: Prentice Hall, 1995.
> An advanced introduction to the fundamental concepts in digital imaging and how they can be applied to real-world problems.

Digital Image Processing: Principles and Applications. Gregory A. Baxes. New York: John Wiley & Sons, 1994.
> An elementary, practical explanation of digital image processing.

Introductory Digital Image Processing: A Remote Sensing Perspective. John R. Jensen. New York: Prentice Hall, 1995.
    Focuses primarily on space imaging for resource management.

Articles

"A Photo Lab on Your Desk." S.G. Thomes. *U.S. News & World Report*, November 25, 1996, 104.
    Digital imaging and personal computing.

"What's New in Digital Imaging." *Fortune* Special Issue, Winter 1997, 188-189.
    A brief overview of recent digital imaging.

Electric Power Generation

BASIC DESCRIPTION

Electric power is provided on a large scale by central utility stations for industrial and residential use. This section covers high-tech methods of generating electricity as well as the utilization of nontraditional energy sources.

The basic operation followed by most power stations is to ignite a fuel such as coal or uranium in order to heat water, and then to use the resultant steam (water vapor) to spin a turbine generator, creating electric current in a cable.

This electric current is then carried by overhead or underground cables to the houses and factories where it will be used to drive motors, provide lighting or heat, and for many other familiar applications.

Note: Three common words ("fuel," "energy," "power") are used throughout this section. The following informal definitions should help to clarify their use:

fuel—a material that can be ignited or otherwise activated to provide useful energy to a system. Examples: wood, coal, gas, uranium.

energy—that condition of a system or device that enables it to perform physical tasks (work). Examples: the pressure in a steam engine or the charge on a battery.

power—the rate at which energy is utilized in a system to perform work. Examples: an airplane may burn a gallon of fuel in a few seconds to move its heavy mass; it's a high-powered system. A small motorcycle might take a couple of hours to consume the same amount of fuel; it's a low-powered system. In addition to the foregoing definition, which is the meaning that most scientists and engineers attach to the word "power," it is also used to refer to a source or means of supplying energy. Example: the power plant.

REPRESENTATIVE APPLICATIONS

General

The use of electricity in the modern home, office, and industrial plant is so deeply embedded in our way of life that it has come to seem like an ever-available natural resource on the order of sunlight and air. It's difficult to envision that only a few generations ago factories and homes were lit by candles and kerosene lamps.

Approximately 90 percent of central station (utility) power generation in the United States is based on burning fossil fuels such as coal, petroleum, and natural gas, or on activating nuclear reactions in uranium. The remaining 10 percent comes from hydropower and a few other minor sources. The future availability of fuels and the impact of their use on the environment are the most serious questions that power companies are now facing.

BACKGROUND

The most widely used system in the United States for generating central station power is the thermal-mechanical method. This method includes both fossil fuel and nuclear reactor plants. Although there are significant engineering differences among these plants, they all adhere to the following basic pattern:

1. A fuel activation unit (such as a furnace for coal, or a reactor for uranium) provides heat to a tank of water.
2. The tank of water produces steam, which is piped to a turbine and forces it to spin. (A turbine is a long shaft with protruding blades or fins, which capture the energy of the steam and cause the shaft to rotate.)
3. The turbine shaft is attached to the rotor (the moving part) of a generator, and when the turbine shaft moves it causes the rotor to spin and activate the generator.

4. The activated generator produces electricity, which is carried from the power plant, either above ground on lines strung from poles or towers, or below ground through buried cables, to the consumers.

The heart of the central power system is the generator. In principle, a generator is a loop of wire rotating in a magnetic field as shown (schematically) below in figure 3.

The principle of the generator was discovered in 1831 by famed British scientist Michael Faraday. He found that as the loop of wire cut across the field lines of the magnet, electric current began to flow in the wire. If the loop of wire was rotated continuously, a continuous flow of current would result.

The discovery of this simple principle eventually led to the effective harnessing of electromagnetic energy (+) and to a radically changed world.

The loop of wire rotating in the magnetic field has evolved into the huge generators that now produce electricity for entire cities.

Virtually all central station power plants produce alternating current (AC). As the loop of wire rotates in the magnetic field it goes through a complete circle. Halfway through each rotation the polarity of the current reverses and the current flows in the opposite direction. Thus the current is continually fluctuating back and forth. If the loop makes a complete rotation 60 times in one second, it is producing 60 cycle AC current. This AC current is the one most commonly supplied in the United States.

While direct current (DC) could be supplied by making a few engineering changes in the generator, AC has been found to be far more practical for high-voltage, long-distance transmission. A particularly important point is that AC voltage can easily be "stepped up" (increased) or "stepped down" (decreased) through the use of transformers. This flexibility allows the power company to adjust voltages to the users' needs. For example, the average household usually has modest voltage requirements (110 or 120 volts), while factories and hospitals may require higher

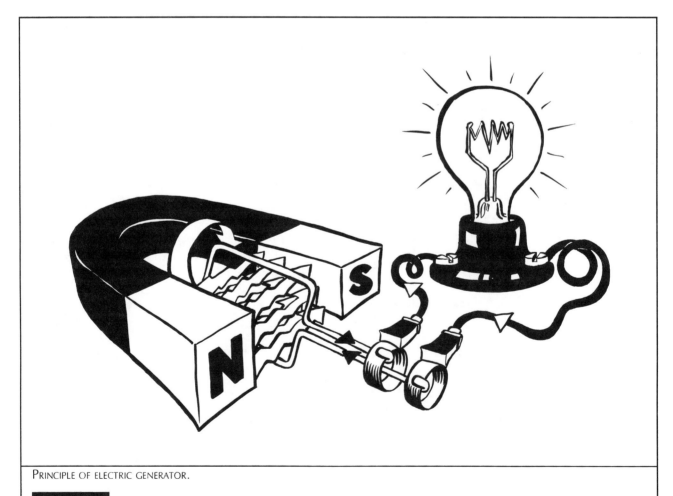

PRINCIPLE OF ELECTRIC GENERATOR.

FIGURE 3

voltages. (Voltage is the measure of the force with which electric charges are "pushed" through a material.)

Since they produce alternating current, central station generators are usually referred to in the industry as "alternators." Also, for practical engineering reasons, the current-carrying coil, consisting of many loops of wire, is generally held stationary, and the magnetic fields are rotated instead.

From these huge generators (or alternators), vast quantities of energy flow out to the cities and countrysides of the world. The part of the process that starts at the generator and ends with the consumers' use of the energy supplied, i.e., the distribution end of the system, is a relatively "clean" process.

Note: Some recent studies have speculated on the adverse health reactions to long-term exposure to low levels of radiation (such as those emitted by electric transmission lines, also referred to as EMFs, which stands for electromagnetic fields). But, generally speaking, this part of the system has not drawn the attention and controversy that the fuel activation segment of the process attracts.

Fuel is needed to supply the energy to make the generator's moving part (the rotor) spin. It is unimportant to the generator whether it is driven by fossil fuels such as coal or oil, or by nuclear fission, solar thermal energy, hydropower, or even discarded tires (as some experimental plants are doing). It does, however, make a great deal of difference to society. Burning fossil fuels causes serious pollution, including the production of "greenhouse gases," which are thought to potentially contribute to global warming. In addition, nuclear reactors are a significant source of radioactive waste.

High-Tech Power Plants

All power plants are largely defined by the fuel that they use. Their physical structure, their operating and safety procedures, and their legal and licensing requirements all reflect the nature of the fuel that drives their alternators.

NUCLEAR POWER

Fission—All existing commercial nuclear power generating facilities utilize the fission process. The word "fission" means the splitting or breaking up of an object; in this case it refers to the splitting of an atomic nucleus.

The fuel employed in the fission process is a mixture of uranium 238 (U-238) and uranium 235 (U-235). The numbers 238 and 235 refer to the atomic weight of the element. U-235 is an "isotope" of ura-

nium 238. That is to say, it has the same chemical properties, but a different weight, and will react differently at the nuclear level. In the natural state, mined uranium ore contains less than one percent uranium and most of that is U-238. Only about 0.7 percent is U-235.

Since only U-235 will undergo fission in a commercial nuclear power plant, the proportion of U-235 in the compound has to be increased to about three to four percent of the total. This somewhat difficult process is referred to as "enrichment."

To begin the fission process, the enriched uranium compound, in the form of pellets encased in metallic tubes, is bombarded with neutrons (electrically neutral nuclear particles) from a strategically positioned antimony-beryllium start-up source. Some of the neutrons strike the U-235 atoms and cause them to break up into a complex of fragments, including additional neutrons.

When these newly released neutrons strike other U-235 nuclei and cause them to split, releasing yet more neutrons, a self-sustaining process, i.e., a chain reaction, has begun. The chain reaction is regulated by the use of control rods (typically boron in steel), which are inserted through predrilled holes directly into the fuel rods. The boron captures free neutrons, so that the process can be regulated and controlled by partial insertion of the control rod, or shut down completely by full insertion of the rod.

The neutrons released in these reactions travel at too high a velocity to permit them to fission the U-235. They need to be slowed down by the use of a "moderator" such as carbon or water.

The disintegration of the uranium nuclei in the chain reaction generates great amounts of energy, so that the particles released (fission products) travel at high speed, colliding with each other and generating heat. To use this heat, the reactor core is immersed in a circulating liquid; the liquid serves the dual objectives of cooling the process and of creating a source of steam to operate the turbine generator. (See figure 4.)

There is no realistic possibility of a nuclear explosion in a commercial reactor, since the process does not employ a sufficiently enriched uranium compound. However, a conventional explosion, (a boiler blast, for example, or a chemical explosion) could occur. This type of accident, while unlikely to be any more severe than it would be in a conventional facility, is vastly complicated by the presence of highly radioactive substances. (Radioactive materials, such as uranium, are substances that spontaneously give off high-speed nuclear particles. Such particles are harmful to human beings and other living creatures.) If the outer protec-

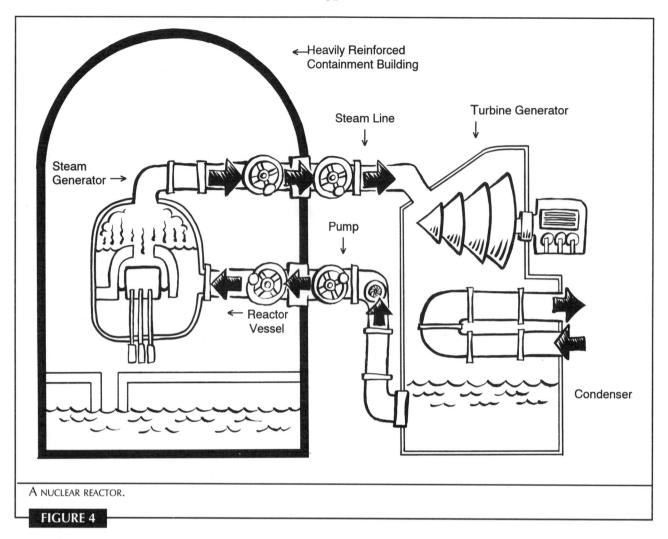

Heavily Reinforced
Containment Building

Steam Line

Turbine Generator

Steam
Generator

Pump

Reactor
Vessel

Condenser

A NUCLEAR REACTOR.

FIGURE 4

tive structure of the plant is breached in such an accident, extremely dangerous concentrations of radioactivity could be released into the atmosphere.

Another hazard is the possibility of runaway overheating in the plant, which could cause a meltdown of the reactor core. This scenario could take place if the cooling system failed. A meltdown would reduce the core to an intensely hot molten mass and could release lethal clouds of radioactivity. While insertion of the neutron-capturing control rods into the fuel core will shut down the chain reaction, the radioactive fission products already produced will continue to generate heat. It is essential, therefore, that the cooling system be reactivated as soon as possible, even after the chain reaction has been stopped.

There is also concern that, in a system failure, the molten core could sink beneath the floor of the plant and penetrate to some depth within the Earth. This occurrence, sometimes referred to as the "China Syndrome," could cause steam explosions and could release additional radioactivity into the air. Underground

water might also become contaminated, which would spread radioactivity over a wide area. After the Chernobyl nuclear power plant partial meltdown in 1986, the molten core was encased in a thick jacket of concrete in an effort to prevent any more contamination (other than the initial radioactive release) from occurring.

In addition to the safety hazards described above, two inherent disadvantages of the fission process must be considered:

1. The uranium fuel that drives the process must be replaced periodically. This process is difficult and hazardous. Even though the fuel is not fully spent, it must be replaced after a certain period of use, because the waste products of the process build up and begin to capture significant quantities of the essential neutrons.

2. The fission process transforms the uranium fuel into a number of highly radioactive frag-

ments. A fully satisfactory method of disposing of radioactive waste products has not yet been developed.

Fast breeder reactors—In the fission process described above, some of the neutrons are captured by U-238, which then decays, through several steps, into plutonium 239 (Pu-239), an element not found in nature. Pu-239 is capable of undergoing fission and, to some extent, does so in the regular fission process.

If the composition of the fuel could be adjusted to the right proportions of U-235, U-238, and Pu-239, it might be possible to produce enough fissionable Pu-239 to keep the chain reaction going without additional fuel. This chain reaction has, in fact, been accomplished (on the pilot plant level) with highly enriched uranium, i.e., 20 to 25 percent U-235 content.

In this process, the neutrons are not slowed by a moderator (since the enriched fuel assures enough fission events), and the U-238 captures many high-speed neutrons and decays into additional Pu-239. Since the Pu-239 is fissionable, new fuel is constantly being produced. Hence the name "fast breeder reactor."

While this process is a great conserver of uranium fuel, its production of, and dependence on, plutonium 239 reduces its appeal. Although these reactors generally don't produce so-called weapons-grade plutonium, any plutonium can be used to make nuclear weapons and could lead to the proliferation of such weapons. It is also highly toxic.

Other reactors—There are other types of nuclear reactors, including light-water reactors (as pressurized-water reactors and boiling-water reactors) and high temperature gas-cooled reactors. Each depends on radioactivity (they all use fission) and the heating of water to produce steam; but each varies in overall design. The details of each are too complex to cover in this text.

Fusion—Fusion is the process by which the Sun generates its energy.

When subjected to tremendously high temperatures (exceeding 50 million degrees Celsius), the nuclei of light elements (such as tritium and deuterium) combine (fuse) to form heavier nuclei, such as that of helium, and, in doing so, release enormous quantities of energy. The energy released in fusion far exceeds that of fission. (Although individual fission events release more energy than individual fusion events, the latter are far more plentiful in a given system.)

Tritium and deuterium are isotopes of hydrogen. Isotopes are variants of an element that have extra neutrons in their nucleus and are therefore heavier than the standard element. Chemically they act like the familiar element but physically their behavior differs.

The fusion process has access to a virtually inexhaustible supply of fuel (hydrogen and its isotopes are plentiful); it produces very little radioactive waste and is self-extinguishing if something goes wrong. But, to date, no sustained, controlled fusion reaction has been achieved, and scientists predict that it will be a matter of decades before fusion power is commercially available.

Scientific research has focused on two primary methods of producing controlled fusion:

1. The plasma method—A plasma is a gas heated to a sufficiently high temperature to strip its constituent atoms of their electrons. (See "The Electromagnetic Spectrum.") If such a plasma could be brought to an even higher temperature, a fusion reaction might take place. One of the difficult problems in this process is to prevent the plasma from dissipating, i.e., to keep it contained. Virtually any physical container that we are familiar with would be destroyed at the millions of degrees Celsius required in the fusion process. However, since the nuclei of the gas have been stripped of their electrons, they are now positively charged and can, at least theoretically, be contained by a magnetic field. This approach is now being pursued in the plasma process.

2. Inertial confinement—In this process a deuterium-tritium fuel mixture is formed into a pellet, onto which an intense beam of energy is focused. The energy beam compresses the pellet with tremendous force, generating intense heat and igniting the fusion process. The inertia (the tendency of a system to maintain its current state) of the process holds the fuel together until fusion energy has been released. The development of intensely powerful lasers (+) to provide the energy beam has brought this process closer to fruition.

Note: In March 1989, two respected researchers caused a major scientific furor when they reportedly started a fusion reaction at room temperature using a simple laboratory procedure. Since that time, "cold fusion," as it is called, has been attempted many times, and there have been several claims of success. But as of this writing, no one has successfully been able to duplicate any of the alleged fusion procedures to satisfy the cold fusion community.

Other Energy Sources

Geothermal—The interior of the Earth is much hotter than the surface. At particular places on the Earth's surface, this heat breaks through to the surface (or just below the surface) in the form of hot water, steam, hot rocks, or the lava of volcanoes. This energy can be used to generate electricity.

Geothermal energy is now being used for central station power generation in several countries. The largest installation in the United States is at The Geysers in California, about 90 miles north of San Francisco. It is capable of generating about 800 megawatts. The plant at The Geysers uses a conventional steam turbine generator, with specialized plumbing to harness the steam coming out of the ground.

The country of Iceland depends significantly on its geothermal sources; about 50 percent of its houses are heated through geothermal piping.

An idea still in the experimental stage involves drilling just over 6 miles into the Earth's surface; at this depth, most rock is about 249 degrees Celsius (480 degrees Fahrenheit) or more. After creating cracks in the rock with explosives, water is then pumped into the hot rock region, heated, then pumped back to the surface to generate electricity. Many difficult barriers currently limit the potential of such systems—especially digging to such depths and the logistics of pumping the water from the surface and back again.

While geothermal energy is a valuable resource and contributes to the world's energy needs, its availability is limited to certain geographical areas. Any large-scale expansion of geothermal power is therefore unlikely.

Hydropower—Hydropower provides less than 10 percent of the United States' power needs. In countries that have plenty of waterfalls along steep gradients, hydroelectric power is much more prevalent. For example, Norway produces 99 percent of its electricity with hydropower; Switzerland produces 75 percent of its power with water. Hydroelectric power does not use any of the fuels listed earlier but instead employs a turbine to capture the energy of water flowing over a waterfall or from the outgates of a dam. However, hydropower is a limited resource that exists, or can be artificially produced, only in certain geographical settings in which there is sufficient running water along a steep gradient. Also, when hydrodams are constructed they can damage or destroy natural land features, and sometimes even force the relocation of people in adjacent areas.

Ocean-based—Tidal power, like that of the wind, is a source of energy that has been used for centuries. It requires the construction of a small dam that accumulates water as the tide comes in and releases it when the tide goes out. A water wheel or turbine is driven by both the outgoing and the incoming water, activating an electric generator. As with geothermal and hydropower, tidal power is confined to particular geographical areas.

Researchers are also investigating the power that might be derived from the motion of waves and ocean temperature differences. One experimental energy plant uses the motion of the waves to compress air within an enclosed chamber. The air passing through to the chamber turns a wind turbine, which in turn, generates the electricity. As the wave retreats, air is sucked out of the chamber, causing the turbine to move in the opposite direction. For the most part, however, these concepts are still on the drawing board.

Biomass—About 13 percent of the world's power comes from biomass, including burning vegetation to create steam and fermenting livestock dung to create flammable gases which are then burned as fuel. Biomass includes wood, crop residues, municipal wastes, and other organic materials that can be converted to heat, process steam, electricity, and transportation fuel. For example, sugar cane is thought to be one of the best energy crops because it grows fast and produces a great deal of biomass. The leaves and tops of the cane plant, and the cane residue left over after the extraction of sugar, are burned to create steam for turbines.

Solar photovoltaic systems—Solar photovoltaic cells are semiconductor (+) devices that generate electric current when exposed to sunlight.

A power plant operating on photovoltaic cells would eliminate all of the furnaces, boilers, and plumbing as well as the turbines and generators that conventional systems require. It would also be nonpolluting and have an inexhaustible source of fuel. However, in spite of the fact that cells have become more efficient and less costly in recent years, engineering and economics do not make them appear to be likely contenders as central station power suppliers. Their best market may turn out to be in individual homes and commercial buildings.

Solar thermal—Most of the Sun's radiation is in the infrared, visible, and ultraviolet segments of the electromagnetic spectrum (+). Utilizing this energy in a large-scale, central power station depends on several variables. One key variable is the weather; a good deal more useful radiation will reach the Earth's surface on a clear day than on a cloudy day. Solar power is not an attractive prospect in a location that has many more cloudy than sunny days.

An installation, even in a very sunny area, will not receive any solar energy at night. It is clear, then, that any large-scale solar thermal facility will require the ability to store energy when it is available. In most of the plans drawn up so far, this storage would be achieved by heating water in sunny periods, storing it in an insulated tank, and using it as needed.

The actual structure of such a facility would include a collector, i.e., a highly polished complex of mirrors that would track the sun and focus its energy into a tank of water. Water (as steam) could be heated in this way to nearly 1500 degrees Celsius (2732 degrees Farenheit). From this point on, in principle, operations would not differ radically from those in a standard fossil fuel plant. Steam would drive a turbine and activate a generator.

In terms of its fuel, such a system would be nonpolluting and inexhaustible. However, the newly developing concerns about the health effects of power line transmission radiation would still be present. Another significant factor is that in any plant layout an enormous amount of land area would be needed because solar energy is so diffused.

Owing to these limitations, solar thermal central station power facilities are likely to be supplementary, rather than primary systems.

Note: A recently completed solar thermal facility in the Mojave Desert is now capable of producing about 274 megawatts and is reportedly a commercial success. It is not a primary source of power but supplements the peak needs of Los Angeles utilities companies. The Luz International Solar Thermal Plant is located about 217 kilometers (135 miles) from Los Angeles.

Wind—Wind is an age-old source of power and has attracted increasing attention in recent years. In principle, the wind tower would capture the energy of the wind and transfer it to a turbine and generator to produce electricity. As with hydropower, no steam generation and associated plumbing is needed. Plans have been drawn up for "wind farms," which would have many wind towers on a single site.

Wind, however, is an intermittent source of power, and it seems unlikely that a system based on its use would ever be more than a minor supplement to a central power plant. In addition, wind towers are somewhat unsightly, require substantial maintenance, and create considerable noise. They have also been known to create problems with certain wildlife, especially birds. For example, bald eagles have a high mortality rate around the Altamont Pass wind turbine farm in California. Studies are now being conducted to suggest turbine-related or habitat modifications that will lower the mortality rate.

THE FUTURE

The "cleanest" sources of power, and those that are both "clean" and renewable, seem to be inherently limited by geography or climate. Thus hydropower, an extremely important source of energy, is limited to areas where there are natural waterfalls or where terrain and social acceptance permit the construction of a dam.

Geothermal energy, wind, and energy from the ocean are also heavily constrained by the realities of geography.

In the United States, biomass energy is often thought of as the "forgotten energy," mainly because of the many local environmental restrictions placed on the burning of biomass and the wastes produced from such burning. But in the early 1980s, several groups were started to encourage greater production of biomass material and more extensive use of biomass burning. For example, the Great Lakes Regional Biomass Energy Program, under contract to the U.S. Department of Energy, supports several states in their search for better biomass energy production.

Solar energy retains significant promise, although much hope for its successful use is in decentralized applications, such as individual residences or vehicles. If the efficiency of solar photovoltaic cells could be brought up sufficiently, and the cost of manufacturing them brought down, they could provide large quantities of electricity without the need for complicated plumbing and massive turbine-generator systems. However, because of the intermittent nature of solar radiation, they are likely to be a supplementary, rather than a primary, producer of electricity in any central station power operation.

For any of these renewable resources to succeed, there must be some encouragement from world governments. For example, in India, the use of wind as energy has gone from virtually nothing to about 820 megawatts—or enough to power about 1 million Indian households—in 1996 mainly through tax incentives. During that same amount of time, wind use in the United States, with few such programs, grew from about 1,484 megawatts to 1,596 megawatts.

One factor that could significantly change the equation for intermittent energy sources would be the development of a process for the large-scale storage of electricity. The inability to store significant quantities of electricity has always been a major headache for utility companies. Much of the equipment that they must have to handle peak load periods stands idle when consumption is down. Batteries do a good job on a small scale, but they can't store the kind of quantities that central stations generate. Elevated water tanks, compressed air, and energized gas storage all seem, at present, to have limited potential. A large-scale, practical, electric storage process has been an elusive goal.

Recent developments in superconductivity (+) have brought this possibility much closer. (This development is discussed in greater detail in the section on superconductivity.) If scientists' efforts in this field are successful, sun and wind energy sources would take on vastly increased importance almost immediately.

The prospects of fission-based nuclear energy remain doubtful. Where fission plants were supplying 20 percent of this nation's electric needs a decade ago, the percentage has now dropped slightly below that. The rest of the world, however, particularly France and Japan, continue to expand their use of this source. For example, in 1950, 72 percent of France's power was generated by hydropower, and none by nuclear plants; by 1985, 65 percent of their power came from nuclear, 23 percent from hydro, and the rest from coal, oil, and gas plants. The switch to nuclear has allowed the country to be less dependent on fossil energy sources from other countries, and to have marked reductions in air pollution. France is not alone: Increasingly pessimistic appraisals of the impact of fossil fuels on the environment, particularly with respect to global warming, have made nuclear energy appear a good deal more attractive to many other countries.

In the United States, many plans have been drawn up for safer fission reactors, including the proposed Process Inherent Ultimately Save (PIUS) reactor, in which a pool of boron in water surrounds the uranium core. If the pressure in the reactor falls signifying trouble, the boron mix, at a higher pressure, floods the core, essentially poisoning the reaction and stopping fission. But construction of such reactors remains at a standstill. The public seems to regard fission as a relatively inefficient process with substantial operational risks, no doubt precipitated by the media coverage of the accidents at Three Mile Island, Pennsylvania, and Chernobyl. Also, the question of disposing of radioactive waste from existing plants has never really been resolved. Although several areas have been chosen as potential nuclear dump sites, protests from communities located near the sites are reported frequently in the media. Fission-based nuclear power is at a crossroads in this country, and at this writing it is difficult to guess which path it will take.

Many experts feel that fusion is the most promising central station power technology on the horizon at present. However, no sustained, controlled fusion chain reaction has yet been achieved. It is possible that recent developments in superconductivity will bring it closer by increasing the availability of smaller, more powerful magnets. Nevertheless, it's likely that we will be well into the twenty-first century before fusion is contributing significantly to our power requirements.

We must also reckon with the possibility that unanticipated problems (such as those that appeared with fission) will develop once the fusion process goes into widespread commercial operation.

FURTHER READING

Books

The Development of Electric Power Transmission. John Al Casazza. New York: IEEE, 1994.
> The evolution and future of electric power systems around the world.

Nuclear Energy: Principles, Practices, and Prospects. David Bodansky. New York: American Institute of Physics, 1996.
> A review of nuclear power development and the major themes attached to fission practices.

Solar Electricity. Tomas Markvart (ed.). New York: John Wiley & Sons, 1994.
> A look at the history, the present, and the promise of solar energy.

Wind Energy in America: A History. Robert W. Righter. Oklahoma: University of Oklahoma Press, 1996.
> A detailed look at the history, development, and production of electricity using the captured power of the wind.

Articles

"Advanced Light-Water Reactors." Michael W. Golay and Neil E. Todreas. *Scientific American*, April 1990, 82-89.
> Improvements in the fission process.

"Catching the Sun to Generate Electricity." I. Peterson. *Science News*, June 15, 1996, 374.
> Current solar power electric generation.

"The Hot Path to Solar Electricity." J. T. Johnson. *Popular Science*, May 1990, 82-85.
> The Luz project in the Mojave Desert.

"Solar Power: Consumers Cash In." C. Brighton. *Popular Science*, January 1997, 32.
> Solar power generation, consumers, and the electric companies.

Fiber Optics

BASIC DESCRIPTION

Fiber optics is the process of transmitting messages or images over long or short distances (a few thousandths of an inch to thousands of miles), by directing beams of light through lengths of very small diameter glass or plastic fibers. The light beams are produced by an electric current in a light-emitting device, such as a laser (+) or a light-emitting diode (LED) (+). The pattern of the light waves forms a code (actually, a translation of the 1s and 0s of the binary [+] code) that carries a message along the fiber optic lines, which act as "light guides." (The light transmission is very fast, literally carrying information at the speed of light.) At the receiving end of the fiber optic line, the light beams are converted back into electric current, and the message is decoded back into the digital 1s and 0s of the original signal. There are a number of advantages to using fiber optics. Light beams are immune to electrical noise and can be carried for longer distances before fading, than can electric current in metal communications cables. To counter the risk of fading, light strengtheners, called repeaters, are often added to refresh the signal in certain applications.

REPRESENTATIVE APPLICATIONS

Air and space vehicles

Aircraft and space vehicles are prolific users of message-carrying wires. Replacing metal wires with glass fibers increases reliability and, by reducing overall weight, helps to lower fuel consumption.

Automobiles

The average car has about 80 light bulbs in it. Replacing any one of them can be a formidable task. General Electric once announced an automotive system that will use only one bright, highly reliable, centralized, nonfilament lamp, which will distribute light to all of the 80 points by fiber optics connections.

Communications

The most important current applications of fiber optics are in the area of telephone communications. Commercial long-distance fiber optics telephone lines are already in place in the United States, Japan, Western Europe, and other technologically advanced areas. AT&T currently transmits 98 percent of its long-distance calls on fiber optics networks. A competitor, US Sprint, routes 100 percent of its calls through fiber optics.

Note: Although it is costly and complicated to install new fiber optics systems where conventional metal cable is still working well, the advantages are so great that replacement is taking place rapidly. In addition to the cost of new installations, technicians must acquire new skills in maintaining fiber optics systems. However, none of these obstacles seem to be powerful enough to slow down the momentum of the move to fiber optics in the industry.

Computers

Direct fiber optics links are being used now in some computer-to-computer hook-ups, such as local area networks (LAN) systems. Increasingly, fiber optics is likely to become the basic transmitting channel in computer networks because of its great data-carrying capacity (also called bandwidth); greater resistance to electromagnetic noise such as radios, motors, or other nearby cables; and speed. (The fiber optic networks operate at 2.5 gigabits per second—as opposed to the 155 megabits per second for copper—a higher capacity that increases the speed of the system.) Also, as interest grows in computers that will use light instead of electric current as their basic processing medium, fiber optics will have a vital role to play.

Industry

Fiber optics devices can be successfully utilized in electrically noisy factory environments without special shielding.

Medicine

Fiber optics devices such as fiberscopes and endoscopes permit doctors to view internal organs and tissues of patients by inserting tiny fiber optics tubes into the body. These devices are increasingly used in medicine for both diagnosis and treatment of various illnesses.

Media

The cable TV industry has realized the advantages of glass cable and is using it in many installations. Networks, accustomed to using satellites for long-distance transmission, are currently experimenting with fiber optics cable hook-ups. In addition to its other advantages, fiber cable is not affected by adverse weather conditions, as are satellites.

BACKGROUND

On December 14, 1988, the first undersea fiber optics cable linking the United States with Europe went into operation. It connects a point in Tuckerton, New Jersey, to a branch point near the European coast. From there, a line runs to Widemouth in England and another to Penmarch in France.

The system was the eighth undersea cable to be laid between Europe and America, but it was the first time that a nonmetallic conductor had been used. Three copper cable systems remain in use; they, along with the communications satellites that span the Atlantic, can carry a total of about 20,000 telephone calls at one time. The new fiber optics cable can, by itself, handle 40,000 calls at a time. (A single pair of optical fibers can transmit over 12,000 telephone conversations simultaneously with little, if any, interference; the same pair of metal wires can only carry 24 conversations.)

In addition to the obvious advantage of greater message-carrying capacity, fiber optics brings many other advantages to the field of communications. It is generally not affected by magnetic interference, atmospheric disturbances, cross talk, short circuits, or heat generation, and it is not a lightning attractor. It offers far more security than metal wire because it is extremely difficult to "tap." (It's virtually impossible to tap without detection.) It does not emit any radiation. (Metal cable does, which is an increasing concern in our densely wired society.)

The signal loss in fiber optics systems (every kind of signal loses power and weakens as it travels along its path) is much lower than in metal systems. For example, the new transatlantic fiber optics cable requires a repeater every 24 to 39 miles. The best undersea metal cable needs a repeater about every 6 miles. In addition to all these advantages, fiber optics systems are smaller and weigh less than comparable metal systems.

THE PROCESS

Signaling with light is a long-established human technique. Using mirrors and lanterns between ships or hills was a familiar practice in past eras. Paul Revere's historic ride depended on an arrangement of lanterns in a distant church steeple. A system of towers in eighteenth-century France, each within sight of the next one, enabled human "transmitters," using light signals, to send messages nearly 150 miles in about 15 minutes. In addition, Alexander Graham Bell invented one of the earliest—but primitive—light-wave communications devices in 1880 called a "photophone."

What makes fiber optics different from these other systems is that none of these methods used "guided" light signals. The signals were simply broadcast through the atmosphere and could easily be interfered with by weather conditions or by physical obstacles. Early in this century, several scientists became interested in the possibility of guiding light through fibers and cylinders, and researchers eventually developed "fiberscopes" that enabled physicians to view internal human organs.

But the development of large-scale guided light communications awaited the arrival of a new generation of light sources and wave guides. In the 1960s, with the development of lasers (+) and light-emitting diodes (LEDs) (+), reliable, controllable light sources were at hand. Industrial researchers now began an intensive effort to produce strands of glass with sufficient purity to serve as wave guides for light signals. (Impure glass would cause the signal to degrade too rapidly.) In 1972, Corning Glass announced that it had produced glass pure enough to carry signals at reasonable loss levels. Rapid advances in technique and commercialization followed these two important developments and brought the technology into the marketplace. Other, more recent developments include the use of cheaper plastic fiber optics for certain conditions, although the dissipation of light energy can approach about 1,000 times that of glass fibers.

At its present stage of development, fiber optics is totally dependent on intermediary electronic devices. In fact, a fiber optics communications link is essentially a guided beam of light traveling between two electronic circuits.

A typical fiber optics transmission would go like this: air vibration (such as the sound of a voice going into a telephone mouthpiece) or direct contact (such as a finger striking a keyboard) causes fluctuations in a small electric current flowing in an adjacent pressure-sensitive circuit. The pattern of fluctuations in the current carries the message of the voice or keyboard to a light-emitting device such as a semiconductor laser or a light-emitting diode (LED) (+). Either of these devices can convert electric current into a light wave. The light wave actually produced is "modulated," i.e., it is reshaped by the message pattern carried to it from the pressure-sensitive circuit. The modulated light wave is then directed, at a very precise angle, into the mouth of a tiny, highly uniform glass or plastic fiber with a diameter smaller than that of a human hair.

This message-carrying fiber is enclosed with other fibers (a "bundle") in a multi-layered jacket of protective material. The glass fiber itself is divided into an inner segment called the "core" and an outer segment called the "cladding." (See figure 5.)

There are three main types of optical cables: Single mode, a single strand of glass about 8 to 10 microns (one micron equals approximately 1/250th the width of a human hair), and in which only one ray of light is transmitted; multimode, made of multiple strands of glass fibers combined into a bundle about 50 to 100 microns in diameter, and in which each strand is capable of carrying a different signal; and plastic optical fibers, a newer, more experimental plastic fiber that has potential to perform similar to a single mode cable, but at a lower cost.

The angle at which the light wave enters the fiber is crucial because the reflective characteristics of the core and the cladding are carefully coordinated to assure maximum reflection of the wave when it strikes the junction between the cladding and the core. We know from experience that if a light wave strikes a smooth reflective surface at the correct (also called critical) angle, it will be almost completely reflected back from the surface at the same angle. (Stated formally: the angle of incidence will equal the angle of reflectance.)

The light wave then proceeds on its path, bouncing off the junction wall between the core and the cladding, all the way to its destination. If the distance to be covered is great enough to weaken the signal excessively, electronic repeaters will be placed along the path to revive it.

At the other end of the fiber, the light signal is captured by a "detector," i.e., some kind of a light-sensitive device such as a photodiode (+). Now the whole process is reversed. The photodiode converts the modulated light wave back into electric current, which fluctuates in accordance with the message pattern carried by the light wave. This current is amplified and used to drive a mechanical device such as a small speaker in a telephone earpiece or a hard-copy printer. A sound resembling the voice of the person at the other end comes out of the telephone earpiece, or a hard copy of the data typed in at the keyboard emerges from the printer. In any case, this is the end of the journey for that light signal. The message has been delivered.

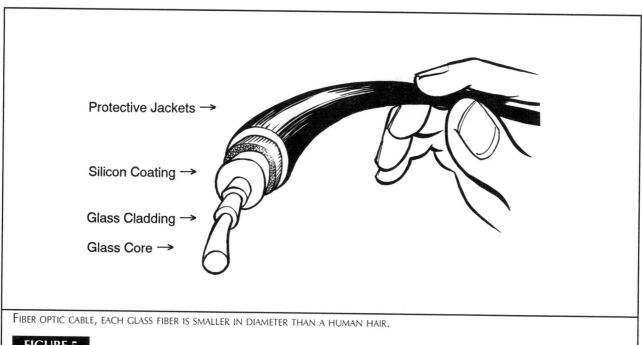

FIBER OPTIC CABLE, EACH GLASS FIBER IS SMALLER IN DIAMETER THAN A HUMAN HAIR.

FIGURE 5

This simplified description gives only a hint of the engineering complexity of modern fiber optics systems. The need to send multiple messages over the same channel at the same time ("multiplexing"), and to send messages in opposite directions over the same channel ("duplexing"), calls for ingenuity in design and painstaking precision in manufacturing.

It should also be borne in mind that fiber optics systems can carry images from one point to another in their original form, i.e., without encoding them into an electric current or light wave. They accomplish this transmission by using one fiber optics strand for each picture element. The medical fiberscope, described above, operates on this principle.

THE FUTURE

The most important and successful application of fiber optics has been in the telecommunications industry. Its impact is so great that it is threatening the preeminence of satellites in long-distance communications. Other than its message-carrying capacity, which dwarfs other systems, one of its greatest assets is its reliability. For example, if projections are accurate, the transatlantic fiber optics cable will require only three repairs over the next 20 years.

Computer companies are now trying to make fiber optics and its wider bandwidth capacity the basic carrying medium for networking (+) projects.

One concept that particularly intrigues planners is directly connecting homes to TV channels, telephone systems, and commercial centers with a single fiber optics link. Such systems would allow users to receive hundreds of TV channels (including the use of high resolution, High-Definition Television [HDTV]), electronic newspapers, and other media; to participate in interactive games; to send messages out of their homes; and to obtain on-demand video films. This set-up would allow people to complain about or praise TV programs, request shopping information from department stores, pay bills, send electronic mail, and perform a multitude of other functions. The overall availability of such systems is rapidly becoming feasible with the increase in the installation of fiber optic cables across the country.

Other observers advise caution with respect to this concept, pointing out that consumers will not necessarily accept these innovations simply because of their technical ingenuity. The picture telephone is often pointed to as a technology that never really found a market. It could be argued, however, that it is not only the lack of consumer acceptance of the videophone that has stopped it from gaining popularity: Even with a lower picture quality and compression (a technique in which repeated bits of data can be removed), videophones still take long periods of time to send images. One model, sending at least 28,800 bits per second, takes about three minutes to send one still image.

Future possibilities for fiber optics in the field of medicine are particularly promising. Fiberscopes and endoscopes continue to play an increasingly important role in diagnosing illnesses. In addition, fiber optics is now being used to actually treat certain disorders by guiding carefully metered beams of light to affected portions of the body.

On the leading edge of this medical technology is an experimental cancer treatment some physicians are trying: These doctors are treating cancerous growths with a technique that consists of having the patient ingest a light-activated substance, such as dihematoporphyrin ether (usually referred to as DHE), which then accumulates in cancerous cells. When a beam of fiber-conducted light is brought to bear on the affected area, the DHE is activated and destroys the cancerous cells.

With respect to telecommunications, the only question remaining is how rapidly glass fibers will replace metallic cables. The process is under way and is gaining momentum. There can be little doubt that fiber optics, at every level of communication, is the medium of the future.

FURTHER READING

Books

Architectural Lighting with Fiberoptics. Gersil N. Kay. New York: W.W. Norton & Co., 1997.
    Applications for architecture using fiber optics.
Fiber Optic Reference Guide: A Practical Guide to the Technology. David R. Goff, Kimberly S. Hansen (ed.), James G. Stewart (ed.). New York: Focal Press, 1996.
    A practical look at fiber optics.
Fundamentals of Optical Fibers. John A. Buck. New York: John Wiley & Sons, 1995.
    An intermediate discussion on the fundamental principles of light propagation and fibers.
Understanding Fiber Optics. Jeff Hecht. New York: H.W. Sams, 1993 (2nd edition).
    Covers theory and applications of fiber optics.

Articles

"Bandwidth, Unlimited." W.W. Gibbs. *Scientific American,* January 1997, 41.
    How wavelength division multiplexing creates optical networks.
"Optical Fibers in Medicine." Abraham Katzir. *Scientific American,* May 1989, 120-25.
    An updating of recent developments in the utilization of fiber optics in medical applications.

Laser

BASIC DESCRIPTION

A laser is a device capable of generating a narrow, uniform beam of concentrated, high-intensity light, made up almost completely of a single pure color. Laser light can be precisely directed over small or large distances (less than a millimeter to thousands of miles). Lasers themselves vary greatly in size; one type is as small as a grain of sand, another as long as a football field. The intensity and diameter of the laser beam can be rigorously controlled. Lasers have become extremely important in communications, manufacturing, medicine, and military applications.

REPRESENTATIVE APPLICATIONS

Commerce/Services

Cashiers in many supermarkets and department stores are now using laser bar code (+) readers to price purchased items and to update inventory records. Libraries, the post office, and banks are also using laser bar code readers to keep more accurate records.

Communications

Laser-generated light beams used as message carriers in fiber optics (+) telephone systems are playing an increasingly important role in communications.

Computers

Many computers are now using laser printers for output. As optical storage (+) memory devices become more common, and light-switching devices come into use, lasers will play a vital role in the new processes. Of particular importance in the future will be computer memories using the principles of holography (+). Such memories, which require lasers to store information and read it out, have tremendous storage capacity. Computer laser printers focus a directional (modulated) laser beam (a beam reshaped to carry a message or image) to a light-sensitive surface, where it leaves its image or message.

Entertainment

Compact disc stereo players using laser readers have become extremely popular and have almost replaced turntables and LP records. Laser-read video discs, called videodisks or laser discs, play an important role in television entertainment systems. The flat, round platters contain recorded images and sounds that can be played back on a television set using a special laser-based player.

Industry

Lasers are being used in cutting, drilling, welding, and annealing operations. Since laser beams can be directed very precisely and their diameter closely controlled, they are particularly effective in these applications. In addition, a laser beam's highly focused light produces a great deal of heat, which is also useful for such applications. Another important use is in the alignment of equipment on the factory floor and in general, in precise measurement for the construction of buildings and close tolerance machines.

Medicine

The tiny, precisely directable beam of energy-intensive light that a laser can provide has already made it a vital new resource in the field of medicine, particularly in surgery. Some inner eye surgery is now performed without the use of cutting instruments by passing a laser directly through the cornea of the eye; other ophthalmologists use lasers to weld damaged retinas in place. Some internal hemorrhaging can be precisely located with an endoscope and then treated by laser light transmitted through another fiber optics (+) device to the area of bleeding. Medical use of lasers has now extended into dermatology, orthopedics, and neurosurgery. Some brain tumors have been treated with lasers. Birthmarks, moles, and tattoos have been removed with laser beams. In general, the ability to direct a tiny beam of highly energetic light to some exact point on or within the body holds great promise for other, as yet unforeseen, medical applications. Nonsurgical use of lasers in medicine includes holography (+), in which resulting holograms are used for diagnostic purposes.

Military

The use of laser radar devices permits much more precise target tracking and identification than can be achieved with conventional radar equipment. In strategic defense planning, the laser could play an important role not only in tracking and identification, but also as a defensive weapon against enemy missiles.

Lasers and laser gyroscopes also play an important role in guiding missiles to targets. Most of the "smart" bombs used in the Gulf War were laser guided. In addition, laser gyroscopes are used to keep ships, airplanes, and spacecraft on course.

Nuclear Power

The laser is also playing a key role in the ongoing search for power from controlled fusion (+). Scientists use the intensely concentrated beams of the laser in their attempts to superheat hydrogen and ignite the fusion process to generate power.

General Research

The laser is increasingly used in research fields such as spectroscopy, i.e., the study of interactions between radiation and matter.

The laser also performs some of its most useful functions by scanning surfaces. The tiny, controllable point of light that it can create is swept systematically across a surface until the entire surface area, or that portion of it that is relevant to whatever determination is to be made, has been covered. The reflected beams, when captured by a monitor and interpreted by some kind of data processor, can reveal salient features of that surface, i.e., irregularities, color, and printed data.

Because of lasers' precision and narrow focus, the beams can be transmitted over great distances with no power loss. For example, laser beams have been bounced off the Moon to determine distance from the Earth to the Moon and to check their movements in relation to each other. Lasers have also been used to detect the slightest movements of the continents, or the slight movement of the crust that may be a precursor to an earthquake.

BACKGROUND

In 1953, Charles H. Townes, an American research physicist, demonstrated a working model of a mecha-

nism that he referred to as a "maser." This device produced coherent, near single-wavelength microwaves. The principles embodied in the maser had been worked out on a theoretical level by Townes and, independently, by two Soviet scientists, N.G. Basov and A.M. Prokhorov. All three later shared the 1964 Nobel Prize in physics for the development of the maser and laser principles.

Following the development of the maser, Townes and others pursued the idea of developing an optical maser, i.e., a device using the same principles as the maser but operating at the frequencies of visible light. However, it was not until 1960 that Theodore H. Maiman, a researcher at the Hughes Research Laboratories in California, demonstrated a working model of an optical maser; it was soon to be referred to as a "laser." (Although it must be noted that the first patent for a laser was given to Columbia University graduate student Gordon Gould, who conceived the idea in 1957.)

Note: Both words, "maser" and "laser," are acronyms. Maser stands for "microwave amplification by stimulated emission of radiation," and laser stands for "light amplification by stimulated emission of radiation."

Maiman's success touched off a tremendous round of research activity and helped to lead the way toward the vast array of laser-dependent devices that now seem to be making their presence felt everywhere in our society.

THE PROCESS

Ordinary light, such as that emitted by a flashlight, is made up of many different colors or, to use optical terminology, many different wavelengths of light. (See "The Electromagnetic Spectrum.") Such ordinary light spreads out and fades before it has traveled any great distance from its source.

In contrast, light produced by a laser can retain a high level of intensity and a relatively undistorted beam shape over great distances. It owes this capability to two important factors:

1. It produces light made up of (almost) a single wavelength. (When many wavelengths are present, they interfere with each other and weaken the light beam.)

2. The light that it produces is coherent; that is to say, its waves are in phase with each other.

We can visualize this process by considering an analogy. If we think of an orchestra made up of instruments of all sizes and shapes, i.e., violins, clarinets, French horns, cymbals, etc., and with each performer playing a different tune at the same time, we have a situation somewhat analogous to that of ordinary light. On the other hand, light produced by a laser can be pictured as the kind of sound that would come from an orchestra made up of only one type of instrument, with everyone playing the same note in unison.

To produce near monochromatic (single wavelength), coherent light, the laser utilizes the light-emitting capability of electrons. Electrons are the negatively charged particles that "orbit" around the atomic nucleus. (See "The Electromagnetic Spectrum.") Electrons are confined to certain specific orbital bands (or clouds as they are sometimes known), and their energy level corresponds to the band that they are currently occupying.

When the filament of an ordinary light bulb is heated by an electric current, its electrons are "excited" (they move to a higher energy orbit), and as they "relax" (return to their original orbit) they emit light. The light coming out of the bulb covers a range of wavelengths and is incoherent, but it provides us with the stable illumination that we are accustomed to and serves our everyday needs quite adequately.

The laser works slightly differently. In general, there are three main parts to a laser: The energy source, such as an electric current, light, or other form of energy; an active medium that is used to absorb the emitted energy for a time, then release it as light; and the optical cavity (or resonator), a structure that holds the active medium.

In a laser, one begins by using a material that will emit a very narrow, predictable band of wavelengths; this material is the medium discussed above. It can be a liquid, a gas, or a solid; the wavelength (and color) of the resulting light will depend on the medium used. For example, a flash of light from a xenon discharge tube can excite atoms in a ruby crystal, creating red laser light. (This type of beam is often used in holograms [+]).When excited by an energy source, the behavior of the medium's electrons will differ in one key respect from those in the filament of a light bulb. Before they relax, or return to their original energy band, the laser electrons move to an intermediate band called the "metastable" state. They tend to remain in this state until nudged out of it by impinging light waves. (This action is the "stimulation" referred to in the acronym that makes up the word "laser.") When this action occurs, the metastable electrons move to a lower energy band and finally to their "ground state," i.e., their original energy band. During this process they emit light at the exact wavelength of the "stimulating" light wave or energy source. The newly emitted light stimulates other excited electrons, which also emit light that is at the same wavelength, and in phase with the stimulating wave. This is the fundamental process that takes place within a laser.

The first laser (that constructed by Theodore Maiman) consisted of a ruby crystal in an optical cavity. Maiman wrapped a flash lamp around the crystal to provide a source of external energy. Two mirrors, one of which was partially transparent, were placed at either end of the ruby crystal. (See figure 6.)

When Maiman powered up the flash unit, it directed an intense pulse of light into the crystal, exciting large numbers of electrons and creating a "population inversion." This phrase is used to describe a situation in which more of the atoms are in an excited state than in a ground state. As the relaxing electrons emitted near monochromatic, coherent light, this light traveled back and forth between the two mirrors, with the flash unit serving as an energy-injecting "pump," amplifying the light emission. As the process continued, some light passed through the semi-transparent mirror, showing the world the first laser light beam.

Ruby lasers are still in use, but lasers have now also been constructed from many other substances. For instance, semiconductor (+) material is widely used in modern lasers. Semiconductor lasers (also called diode lasers) can be extremely small (less than a millimeter in any dimension), and they are "pumped" by electric current. The ends of the semiconductor are polished to form the necessary reflectors. Some of the semiconductor lasers are "tunable," i.e., their frequency can be varied by changing the current passing through them. They are highly efficient and have found important applications in fiber optics (+) communications systems and many other areas.

Other types of lasers include solid-state lasers, gas lasers, and dye lasers. The first laser constructed was a solid-state laser. These lasers use a solid rod, such as crystals or glass, as the active medium. They are used mostly in industry to drill and weld material. Gas lasers use a mixture of gas, or one gas as an active medium. They are used mostly in communications, holography, printing, scanning, and eye surgery. Dye lasers use a dye as the active medium, usually dissolved in a liquid. These lasers are mostly used for researching how colors are absorbed by various materials.

Lasers are also divided into pulse lasers, or those that generate light in pulses and produce energy in fractions of a second (they are often used in the

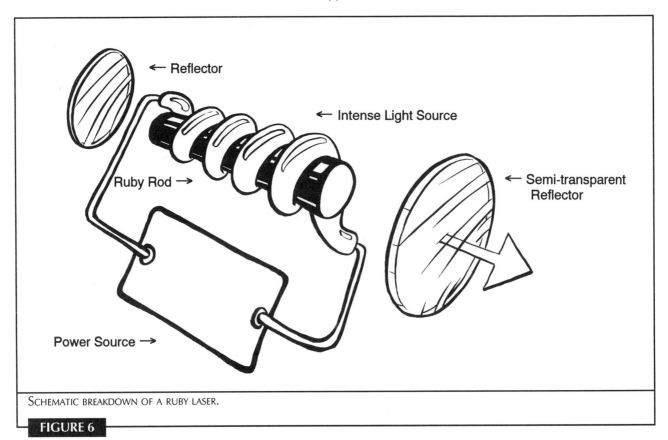

Schematic breakdown of a ruby laser.

FIGURE 6

medical field, such as removing delicate tissues); and continuous-wave lasers, or those that produce a continuous beam (they are best for cutting objects, such as hard metals).

THE FUTURE

Research is continually being carried out in a search for more effective lasers. Many researchers believe that the free electron laser (FEL) is a particularly promising device because it is capable of operating efficiently at high power levels and within a wide range of wavelengths. It is known as the free electron laser because it does not involve the use of electrons associated with atoms but depends, instead, on electron accelerators as a source of free electrons. A stream of electrons, at high speed, is injected from the accelerator into a tunnel equipped with powerful magnets called "wigglers," which change their polarity according to a preset pattern. These wigglers cause the electrons to swerve and, in doing so, to emit light. The light emitted from the electrons amplifies a laser beam that is injected into the tunnel from an external source.

A very high intensity beam of coherent light emerges from the system. The two great advantages of FELs are their power and tunability (the ability to change wavelengths). They are being considered for heating plasmas in the effort to create chain-reaction fusion (+).

Still another advancing area of research is that of the X-ray laser. Because of the very short wavelength of X-rays (see "The Electromagnetic Spectrum"), it should be possible to use them to study very small biological entities that still elude currently available probes. While the principle of operation of an X-ray laser is essentially identical to that of a visible light laser, the need for huge infusions of energy to activate the X-ray device creates formidable engineering problems, which thus far have prevented practical applications from being developed.

Within medicine, the precision and versatility of the laser continues to open doors to new procedures, particularly eye surgery and the treatment of tumors.

In general, applications for lasers continue to proliferate so rapidly that it is difficult to keep pace with them. There can be no doubt that this trend will continue.

FURTHER READING

Books

Laser Fundamentals. William T. Silfvast. England: Cambridge University Press, 1996.

A introduction to the physical and engineering principles of laser operation and design.

Understanding Lasers: An Entry-Level Guide. Jeff Hecht. New York: IEEE, 1994 (2nd edition).

Entry-level introduction to laser technology.

Articles

"Eye Surgery? Take a Close Look." R. Korman. *Business Week*, January 27, 1997, 102-103.

Current applications of lasers in eye surgery.

"Free-Electron Lasers: Present Status and Future Prospects." Kwang-Je Kim and Andrew Sessler. *Science*, October 5, 1990, 88-93.

Current status of this new technology.

"Graffiti Busters." P. Tyson. *Technology Review*, August/September 1996, 16.

Using lasers to get rid of graffiti.

Machine Vision

BASIC DESCRIPTION

Machine vision is the process by which a computer-driven device optically senses external objects and, from its analysis of the sensed data, infers accurate information about those objects. The sensing device is usually a video camera that is linked to a computer that digitizes (see "Digital Image Processing") the captured images and analyzes them. Before truly advanced robots (+) are produced, machine vision systems with very high-level capabilities will have to be developed.

REPRESENTATIVE APPLICATIONS

Industry

Machine vision devices are in wide use in industry. Generally speaking, they are very limited, very specific devices that perform repetitive tasks in a controlled environment.

Typical tasks include checking parts for correct size or shape, checking the dimensions of automobile assemblies, and inspecting integrated circuits and other parts. The machine vision devices handle these tasks very well; there is no fatigue factor, and the devices are able to discover defects that would easily escape the human eye. Their widespread use has enhanced quality standards and has reduced costs.

Vision systems are also being used to enable stationary robots to locate seams for welding and perform other precise positioning tasks.

Medicine

In diagnosing and treating ailments, the medical community uses images in many forms, including X-ray plates, microscope slides, and specialized photographs. Interpreting these images can be a difficult and time-consuming process, requiring the services of highly qualified people. On an experimental level, machine vision devices are being used to assist in this function. They are also used to screen chromosome slides and cancer smears.

Military

The interpretation of aerial photographs is another extremely complex, time-intensive task that requires the attention of trained personnel. The use of machine vision to analyze and interpret these photographs shows considerable promise.

Robot (+) Navigation

The robotics industry, not to mention the space agencies, anxiously await further development of devices that will provide practical vision systems for mobile robots. For self-navigating robots, a reasonable level of visual competence would represent a tremendous step forward. Even for stationary robots, the addition of sophisticated vision systems to their array of sensory capabilities would make them far more versatile than they are today.

BACKGROUND

As computers and robots have become increasingly capable, the concept of providing them with some ability to "see" objects around them has become more important.

Early work on computer vision, dating back to 1967, was done by Marvin Minsky, Seymour Papert, and Thomas Binford at MIT. While they began their work with great optimism and the expectation of quick results, the problem soon proved to be far more complex and intractable than they had imagined.

Considerable progress has been made since those early days, and many commercial machine vision systems are now in operation. There is still general agreement, however, that human vision is a far more complex process than the early researchers had imagined.

Machine vision systems are divided into two broad categories: special purpose (or "dedicated") systems and general purpose systems. The great majority of systems currently in use are dedicated systems.

Dedicated Machine Vision Systems

In a typical dedicated machine vision system, a camera, similar to those used in the television industry, is linked to a computer, which converts images captured by the camera into numbers (see "Digital Image Processing") and analyzes them. Special lighting and precise positioning of the object under examination are usually necessary to get an adequate image.

The computer makes a comparison between the captured image of the object (represented numerically) and the standard image (held in its memory). It then stores the result for later analysis or reports it immediately to a control processor. The latter may generate a warning signal immediately if a defective part is found, i.e., the captured image does not match the standard. The items under inspection might be a valve, a ball bearing, an integrated circuit, or some other part.

The dedicated machine vision unit performs this kind of job very well, and the use of such devices is expanding rapidly. The microelectronics industry, with its tiny, highly complex chips (+), is a particularly heavy user of these systems. The sensors (cameras) can resolve detail with far greater accuracy than the human eye. Machine vision systems have become all but indispensable in Silicon Valley.

However, such devices can perform only a very limited range of tasks and must operate within a constrained environment. Lighting must be just right, the position of the object under examination must not vary, and the angle of the camera must be fixed. The operations are repetitive and limited in scope.

Medical applications and aerial map interpretation are also situations where the operation can be conducted under closely controlled conditions.

It is when the computer vision device leaves the familiar world of controlled environments and attempts to operate under real-time conditions in the field that a whole new collection of problems appears.

The General Purpose Vision System

The disorderly, confused, ever-changing, bright-gray-dark, multi-colored world that a self-navigating robot might encounter presents a vastly greater challenge than does checking parts on a production line.

If we consider what human vision accomplishes, we can begin to appreciate the immense difficulties that confront designers of vision systems for autonomous robots or any other devices that hope to interpret optically sensed data for real-time interaction. Our "sensors" (eyes), which capture the "raw data" (light waves) depicting our surroundings at any moment, form the "front end" of a highly organized, complex, internal information-processing system.

We know a nerve network evaluates and transmits processed visual signals to the brain to form images. But the images formed would be useless if they could not be referenced against the vast "knowledge base" that is stored, in a manner not fully understood, in the brain. This knowledge base, accumulated through the experiences of a lifetime, contains not only information but retrieval rules and classifiers that permit speedy (and below conscious level) identification of objects and actions within visual range.

We do not really "see" everything that our eyes sense. If we are driving slowly along a suburban street, looking for a particular house number, we may not "see" trees, shrubs, and colorful flowers, which are irrelevant to our immediate mission. (On the other hand, if we have a special interest in flowers, we will probably "see" them anyway, in spite of their irrelevance.) We say that certain things don't "catch our eye." But what really happens is that the mind filters out the irrelevant details. So the eye-mind combination is able to focus on, or "zoom in" on, those features in our surroundings that interest us and ignore most of the rest.

Conversely, we can "see" a great many things that our eyes do not observe. When we see a flat surface with windows on it, we may "see" a rectangular building extending inward a certain distance from the street. If we see a squirrel's head protruding from behind a tree, we may also "see" four little feet and a bushy tail. Whether or not we actually form an image of such features, any actions that we may take with respect to the observed object will certainly proceed in the firm belief that these characteristics are present. The machine, on the other hand, will have a problem just in determining that the squirrel's head is not part of the tree. (See figure 7.)

We are able to distinguish among objects with such ease that it is difficult for us to understand why this area presents such great difficulty for machine vision designers. We have no trouble telling where the box on our desk ends and the desk begins. We rarely mistake shadows for separate objects. We do not suddenly misidentify an object because the lighting around it is changed. We do sometimes, but rarely, have trouble telling whether an object is small or just far away. We can sometimes identify particular individuals at a distance, just from the way they walk.

The vision-mind complex triggers many physical actions that take place without apparent conscious effort and within a very short time interval. The sight

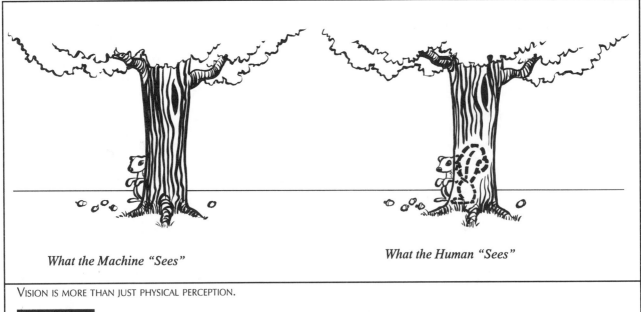

What the Machine "Sees"   *What the Human "Sees"*

VISION IS MORE THAN JUST PHYSICAL PERCEPTION.

FIGURE 7

of a patrol car on the expressway will cause many drivers to glance reflexively at their speedometers. The sight of blood may cause some people to faint. A tennis player at the net can sometimes return a ball that was hit with great speed from only a few feet away. So the vision-mind network can apparently send signals directly to the body's muscles and metabolic regulators without any thoughtful intervention on the part of the individual.

The present state of computer and robotic technology does not begin to approach the capabilities of the human vision-mind-body complex. The most successful experimental systems, to date, have been those that have confined themselves to very limited objectives. These systems generally work along the following lines:

1. Optical sensing of a scene.
2. Division of the scene into objects.
3. Analysis of the features of the objects.
4. Identification of the objects.
5. Reference of the analysis to the device's knowledge base.
6. Generation of signals for appropriate action.

Such laboratory systems have been successful in the identification and, in some instances, the physical manipulation of geometric shapes in the form of cylinders and blocks. A few systems have been successful in recognizing human faces. Progress in this area is increasing.

THE FUTURE

To date, the key to the successful use of machine vision systems has been to confine them to limited tasks within a controlled environment. Researchers refer to this as "domain restriction." The effort to reach more ambitious objectives continues, however. The recent revival of connectionism (the use of hundreds, or more, small interconnected processors; see "Artificial Intelligence") has opened the door to systems that more closely resemble the operations of biological vision networks.

More advanced hardware is certainly going to be needed for more sophisticated machine vision systems. For example, the cameras currently in use are still heavily influenced by their TV background and are generally more suitable for producing images for human eyes than for computer analysis. Connectionist methods will require substantial departure from conventional computer design, which uses sequential processing, to massive networks of strategically connected low-capability processors operating in parallel.

The construction of adequate knowledge bases, general enough to give the system scope but specific enough to provide depth in critical areas, remains a daunting challenge.

Research continues in a effort to better understand the process of human vision and its integration with the mind. Perhaps some significant breakthrough will have to be made in the understanding of this process before a true high-capability general purpose vision system can be developed.

FURTHER READING

Books

Intelligence: The Eye, The Brain and the Computer. Martin A. Fischler, Oscar Firschein. Reading, MA: Addison-Wesley, 1987.
Includes a long section on computer vision.
Machine Vision. Rangachar Kasturi, Brian G. Schunck, Ramesh C. Jain. New York: McGraw Hill Text, 1995.
A technical overview of machine vision.

Articles

"Image Representations for Visual Learning." T. Poggio. *Science*, June 28, 1996, 1905-1090.
An update on visual learning.
"Team Targets Greater Imaging Capabilities." J.T. McKenna. *Aviation Week & Space Technology*, September 2, 1996, 173.
Offers a look at the parallel algebraic logic chip effort.

Medical Imaging

BASIC DESCRIPTION

Medical imaging is the process of creating visual depictions of internal organs, tissues, and biological functions in living human beings.

Important imaging techniques that utilize high technology include CT scans, DSA, PET, SPECT, MRI, and Ultrasound.

CT and DSA use X-rays and require that the patient take contrast agents internally. PET and SPECT create an image by computer analysis of the activity of particles given off by a radioactive compound that the patient takes internally. MRI uses electromagnetic waves (+), much below the frequency of X-rays, to create an image by computer analysis of the behavior of magnetically aligned, radio-deflected hydrogen atoms in the body. Ultrasound uses computer analysis of sound waves echoing from internal parts of the human body to create images.

PET and MRI are particularly costly techniques because they require very elaborate equipment. PET requires on-site use of a cyclotron (particle accelerator) to produce the short-lived radioactive agents the patients take. MRI uses extremely powerful, helium-cooled superconductive magnets. (See "Superconductivity.")

REPRESENTATIVE APPLICATIONS

CT or CAT (Computered Tomography or Computered Axial Tomography; both acronyms are used)

It is used to detect internal bleeding, brain pathology, head injuries, tumors, extent of abdominal injuries, complications of pancreatitis, bone deformities, and osteoporosis. Use of CT has effectively done away with a number of exploratory surgical procedures. CT scans have also been used on mummies to examine the details of the body inside without removing the delicate material holding the body.

Another form of CT is High Resolution Tomography, in which the internal structure, for example, of bone, is imaged at a higher resolution than conventional CT scans. Most slice images (or the cross-section of the bone) have a resolution of about 0.5 millimeters; some researchers claim that they have reached a resolution of about 225 micrometers.

DSA (Digital Subtraction Angiography)

DSA is used to image the structure of blood vessels, particularly the carotid arteries leading to the brain.

MRI (Magnetic Resonance Imaging)

MRI is used to detect a multitude of problems such as those inherent in brain pathology, cardiovascular disorders, congenital heart disease, and pelvic problems. It is not effective with bone problems or where rapid bodily action is present, although it is used to evaluate the spinal cord and vertebrae.

PET (Positron Emission Tomography)

PET is used to evaluate body function such as blood flow and volume (particularly in the brain) and other metabolic processes. Parkinson's and Alzheimer's diseases, epilepsy, strokes, and schizophrenia are among the serious disorders that can be studied with this technique. It is also being used to study normal body function, particularly in the brain.

SPECT (Single Photon Emission Computered Tomography)

SPECT is used to evaluate body function such as cerebral blood flow, activity in lungs, liver, spleen, and bones; and to assess damage in heart attack patients. Important work has been done with Alzheimer's disease.

Ultrasound

Ultrasound is used most commonly in obstetrical evaluations. It is also used to measure blood vessel flow and heart valve function, to detect abdominal abnormalities, gallstones, aneurysms, cysts, and liver and kidney disorders. It is not effective with lungs, bones, or the brain (except in newborn children). Some patients, such as those who are obese, or who have scar tissue in the area to be examined, are not good subjects for this technique.

BACKGROUND

In 1895, Wilhelm Konrad Roentgen, a German physicist, accidentally discovered X-rays while experimenting with a Crookes tube (a device capable of generating a stream of electrons [+]). Roentgen discovered that these mysterious rays were given off when high-speed electrons from the Crookes tube were directed against a solid "target." The heavier the target, the denser the X-rays produced.

The nature of these rays was not fully understood until 1911 when Max Theodore Felix von Laue (also a German physicist) proved that X-rays belonged to the high-energy, short-wavelength region of the electromagnetic spectrum (+).

Roentgen received the Nobel Prize in physics in 1901, and Laue received the same award in 1914.

The ability of X-rays to penetrate solid matter led to their widespread use in medical diagnosis and in engineering evaluation of materials.

X-ray techniques remain an extremely important diagnostic tool in medicine. In fact, about 80 percent of medical imaging is still done with conventional X-rays.

During World War II, interest developed in using radioactive materials in medical treatment and diagnosis. (Radioactive materials are substances that spontaneously emit nuclear particles or radiation.) In the early 1950s, nuclear techniques came into wider use, and they were the forerunners of the modern SPECT and PET procedures.

While Ultrasound (high-frequency sound) has been utilized for many years in a host of applications, it was first used to examine an unborn child by Ian Donald, a Scottish physician, in 1958. Since that time, with the help of the computer, it has become an increasingly important medical diagnostic resource.

In the early 1970s, an important new technique that combined X-rays, high-sensitivity sensors, and computer processing was introduced. It was referred to as computed axial tomographic scanning (soon shortened to CAT scan), and it quickly became a vital diagnostic tool for physicians.

For many years, a research technique known as nuclear magnetic resonance (NMR) was used in laboratories around the world to analyze the interactions of matter and electromagnetic radiation. In 1971, a paper appeared in *Science* which presented data indicating that cancerous and normal tissue in human beings could be distinguished on the basis of an NMR signal. A 1973 paper in *Nature* outlined an actual approach to NMR imaging in human bodies. The use of this new technique became widespread in the 1980s under the name magnetic resonance imaging (MRI). It has now become a major diagnostic tool for physicians.

CLASSIFYING THE IMAGING TECHNIQUES

All of the high-tech imaging procedures listed earlier make use of the computer to analyze the data acquired from examining a patient. Starting from this common point, there are several ways to subdivide or classify the procedures. From the patient's point of view it is important to know (1) whether the procedure will require swallowing, inhaling, or injecting some foreign material into the body, and (2) whether or not it will use ionizing radiation.

With respect to point 1, some procedures require the ingesting of contrast agents by the patient to help the attending physicians get a clear picture of the affected area or organ. The introduction of any foreign substance into the body creates the possibility of allergic reactions or other adverse effects. While such reactions are rare in practice, the risk must be considered. The requirements of the six procedures with respect to contrast agents are shown below:

PET, SPECT	The patient must ingest, or be injected with, a radioactive substance that will emit particles as it moves through the area of the body under study.
DSA, CT	The patient must ingest, or be injected with, a "contrast" medium such as iodine, which will clearly delineate the area under study.
MRI	On occasion, a paramagnetic substance such as gadolinium may be given to the patient. A paramagnetic substance is one that magnetizes parallel to the magnetic field in which it is placed. In MRI this property enhances the image.
Ultrasound	Does not use contrast media or radioactive substances.

With respect to point 2, it is important to know whether the procedure uses ionizing radiation. Ionizing radiation is that portion of the electromagnetic spectrum (+) that includes short wavelength, high-energy waves. Excessive exposure to these waves can damage a biological system. It has been a paramount goal of medical practice to keep such exposure to the minimum necessary for successful diagnosis. Fortunately, technological advances have led to ever-shorter exposure times. In medical imaging practice, X-rays and gamma rays (+) are the two types of ionizing radiation that are likely to be encountered. Requirements with respect to ionizing radiation are shown below:

CT
DSA
PET
SPECT
} All of these processes require the use of ionizing radiation.

MRI
Ultrasound
} These processes do not use ionizing radiation.

PROCEDURES

CT (or CAT) Scans—In the conventional X-ray process, a stream of X-rays is sent through the body, emerges from the opposite side, and immediately strikes a sheet of photographic film.

Massive structures in the body such as bones and cartilage allow fewer X-rays to pass through than do thin tissues.

When the film is developed, it will show (as would any photographic negative) clear areas where there were dense sectors (such as bone) blocking the X-rays, and dark areas where there were relatively transparent tissues allowing most of the X-rays to get through. And, of course, there will be shades of gray between the darkest and lightest areas.

Although an expert can read these shadowy images reasonably well, it is not an easy task. Because the X-rays have passed through the entire width of the body, bones, organs, and tissues are all superimposed on each other, and the patterns can be very confusing. In addition, a great deal of subtle detail may be lost.

In 1979, Allan MacLeod Cormack, an American physicist, and Godfrey Newbold Hounsfield, a British electronics engineer, shared the Nobel Prize for developing a new method of using X-rays. The new process had first been used clinically in 1973. It was referred to as "computed axial tomography." This term was soon shortened to CAT and, later, shortened again to CT.

The development of the CT scanner overcame many of the problems inherent in the conventional X-ray process. In this procedure, the patient lies still while an X-ray tube, supported by a large doughnut-shaped frame, rotates rapidly around the portion of the body under study, sending very quick bursts of X-ray energy through a cross section of the patient's body. These short bursts, which last less than a second, emerge from the tube in the form of narrow (less than half an inch), fan-shaped beams.

As a beam emerges from the body, it strikes a series of tiny, X-ray sensitive electronic devices ("detectors") that have been carefully positioned on the same doughnut-shaped frame, but on the opposite side from the X-ray tube. The detectors rotate in "sync" with the tube.

The detectors record the strength or weakness of the emerging X-ray beams (which will depend on whether they have passed through relatively transparent tissue or dense bone) just as conventional film would do, but they are much more sensitive (hence the very short exposures) and much more precise.

The ring-shaped frame continues to rotate around the patient until it has covered about a half circle (180 degrees). During this process it may have recorded 10 to 30 images from as many different angles but all in the same narrow plane. The newest machines, which may use hundreds of detectors, can complete a scan in as little as two to three seconds.

Since the ring-frame has taken images all around the body of the patient, the same internal organs or tissues have been "viewed" from many different angles.

The varying intensities of the beams striking the detectors are assigned numbers (see "Digital Image Processing") by the computer.

The computer now processes and analyzes the vast collection of "intensity" numbers that it has acquired through the detectors and then reconciles the different scans of the same body feature taken from different angles. Finally, it constructs a remarkably accurate and detailed visual image of a cross section or "slice" of the patient's body. If necessary, additional "slices" can be taken at varying positions in the area of interest.

The images are displayed on a video screen and printed out to obtain a hard copy. The images also can be stored in computer memory. (See figure 8.)

It is estimated that a CT scan will subject the patient to less radiation exposure than a corresponding conventional X-ray examination and that it will provide at least 20 times as much detail.

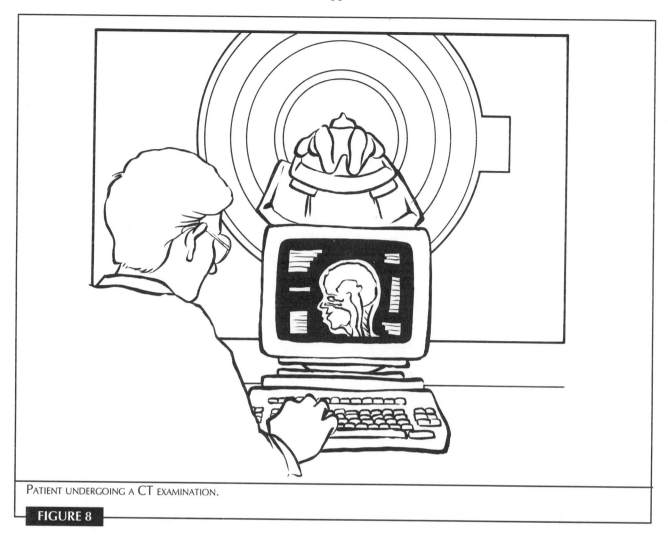

PATIENT UNDERGOING A CT EXAMINATION.

FIGURE 8

CT techniques also make it possible to explore areas that are inaccessible to conventional X-rays, such as certain areas of the brain. Where conventional X-rays detect density differences of 10 percent, CT detects differences down to 0.5 percent, thus allowing smaller potential trouble spots to be found, such as small tumors. In addition, the use of CT, because it is noninvasive, has greatly reduced the need for the more risky angiograms.

DSA (Digital Subtraction Angiography)

The prefix "angio" is derived from the Greek word for vessel and, in medical usage, generally refers to blood vessels.

In this procedure, a conventional X-ray picture is taken of the area of interest, which, for example, might be a carotid artery leading to the brain. This image has the usual drawbacks of conventional X-ray procedure, i.e., nearby organs and bones may obstruct a clear view of the artery. The image is digitized (see "Digital Image Processing") and stored in a computer.

The patient then receives an intravenous injection of a contrast medium such as an iodine compound. When this substance arrives at the targeted artery, another X-ray is taken. This image is also digitized, and then the two images are superimposed on each other (not literally; this is a digital operation within the computer).

Those picture elements unaffected by the contrast agent in the second image are matched with their corresponding elements in the first image and are cancelled out (subtracted) from the composite image. The physician now has a clear view of the artery, outlined with the contrast agent, and free of nonrelevant visual obstructions.

In addition to the advantage of providing the physician with a very clear image, this technique permits the use of much smaller quantities of contrast agents than in conventional techniques.

MRI (Magnetic Resonance Imaging)—Although MRI (formerly called Nuclear Magnetic Resonance [NMR] Spectroscopy) has only been in extensive clinical use since 1981, it has had tremendous impact on the field of medical imaging. It is a radical departure from conventional X-ray technique and makes no use of ionizing radiation.

MRI creates images by analyzing signals given off by certain atomic nuclei within the body. The signals analyzed are those emitted by the nuclei after they have been aligned by an external magnetic field (this alignment is done so that they all have the same orientation), and then deflected by radio frequency waves.

MRI techniques can be employed only with atomic nuclei that possess their own tiny magnetic field. Such nuclei have either an odd mass number or an even number made up of odd numbers of protons and neutrons. (See "The Electromagnetic Spectrum.") As a practical matter, MRI focuses on hydrogen nuclei because of their magnetic field and their abundance in the human body.

Under average conditions, atomic nuclei within human tissue are spinning on randomly oriented axes. When a sufficiently powerful magnetic field is imposed (by placing the patient within a magnetized enclosure), those nuclei possessing magnetism will line up their axes of spin parallel to that of the imposed field.

Radio pulses are then directed at the aligned nuclei, and resonance is created in the atoms (resonance: an echoing level of excitement and activity similar to that caused by one tuning fork in another).

As the nuclei relax (return to their former state) after the radio pulse, they give off characteristic signals, which are captured by a sensitive receiver and recorded on a computer. The captured signals are then used by the computer to construct an image (see "Digital Image Processing") of the area under study.

MRI signals convey a large amount of information because nuclear relaxation times actually provide two readings; the first is the time taken for complete deflection by the radio wave, and the second, the time taken to return to the aligned position. (It is known that an increase in these times occurs when there is cancerous growth in tissue.) Another source of information in MRI follows from the fact that MRI scans can be taken at any angle through the body. CT is limited to horizontal slices through the body because its detectors are mounted on a rigid frame encircling the patient.

The magnets used in the MRI process are so powerful that special precautions must be observed within their range. Any loose metal object can be turned into a dangerous projectile, and unprotected electronic devices may be disabled. The entrance to the MRI room is usually guarded by a metal detector.

MRI is still a very costly procedure, particularly because its powerful magnets require superconductor (+) cooling. In addition, it is still a relatively slow process, requiring the patient to remain still for an extended period in a relatively small enclosed area. (One physician reported that as many as 10 percent of patients are unable to undergo MRI because of feelings of claustrophobia. Newer machines referred to as "Open" MRI have been designed to lessen this anxiety somewhat.) However, the freedom of the process from ionizing radiation represents an advantage over older techniques. In addition MRI images provide greater detail in some situations.

PET (Positron Emission Tomography)—In the PET process, the patient inhales, ingests, or is injected with a substance that has been "labeled" with a radioactive isotope (+). The word "labeling," in this context, means mixing or combining the radioactive material with a substance that occurs naturally in the body, or with an "analog" of such a substance. (In the sense in which it is used here, an "analog" is a synthetic material that will behave in most ways like the natural body substance.)

The purpose of "labeling" the analog or actual substance with a radioactive agent is to be able to follow its path through, and its functional interaction with, the selected organ or area of the body. As the radioactive agent gives off particles, the presence of these particles is "detected" and analyzed by a computer, and an image is constructed of the area where the substance is present. The substance to be labeled, whether an analog or an actual body substance, is chosen because it is known that it will go to the organ under study and will behave predictably when it gets there.

Some radioactive isotopes commonly used as "labels" are oxygen 15, nitrogen 13, and carbon 11. These isotopes emit particles called "positrons," which have all the characteristics of an electron (+) but bear an opposite (positive) charge. Positrons are among the entities that physicists refer to as "antimatter." The key characteristic of "antimatter," as far as its use in PET imaging is concerned, is that, when it comes into contact with its conventional-matter counterpart (an electron), the two will annihilate each other and convert the energy they possessed into gamma rays (+). It is the behavior of these gamma rays that will be detected and analyzed by the computer.

This predictable behavior of matter and antimatter forms the basic principle of operation of PET technology. It is a puzzling fact (at least to physicists) that antimatter seems to be so rare in the universe. (It's a fortunate fact, however, because if matter and antimatter were present in nearly equal proportions, their compulsion to annihilate each other would not bode well for the survival of our familiar environment.)

In an actual clinical situation, the radioactive compound is administered and the patient is placed inside a large ring-shaped frame containing an array of tiny detectors. (This frame resembles the one used in CT scans, but, of course, has no X-ray tube. See figure 9.)

Since positrons (being antimatter) are rare and electrons are commonplace, emitted positrons do not travel far from their isotopes (a tiny fraction of an inch at most) before they encounter an electron. When this meeting takes place, the two particles disappear and are replaced by two gamma rays that travel away from each other, in almost exactly opposite directions.

This critical event in PET imaging is what has made it such an important imaging technology. Because two oppositely directed gamma rays follow each positron-electron encounter, each ray can be captured by a separate detector in the large ring-frame surrounding the patient. Thus there are two readings for each particle emission. If for some reason only one gamma ray is detected after an annihilation event, it is disregarded. But the fact that most of the time there are

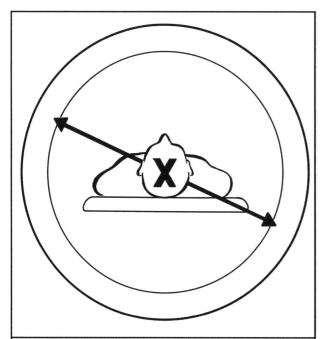

IN A PET ANALYSIS, GAMMA RAYS GENERATED BY A POSITRON-ELECTRON ANNIHILATION TRAVEL IN OPPOSITE DIRECTIONS AND STRIKE DETECTORS IN THE RING AROUND THE PATIENT. THOUSANDS OF SUCH READINGS WILL PERMIT THE COMPUTER TO CONSTRUCT AN IMAGE OF THE AREA UNDER STUDY.

FIGURE 9

two readings means that when all the readings are accumulated, reconciled, and analyzed by the computer, the position of the "label," and therefore the position and flow of the "labeled" compound, can be determined with great accuracy.

By observing the activity levels of labeled substances in the brain when the subject is asked, for example, to respond to thought-provoking questions, investigators are able to pinpoint those areas of the brain that become active as the person ponders the answers. The buildup of this type of information will help to determine, with reasonable certainty, where certain analytical functions in the brain take place. The same techniques can be used to find out where activity is taking place in the brain as people react to things that they see and hear.

Because of its sensitive evaluation of biological function, it is likely that PET will continue to be used extensively with normal volunteer subjects, in order to enhance medical knowledge of the still only dimly understood functions of most areas of the brain. Such research not only benefits medicine, but it also is important to computer scientists working on artificial intelligence (+), who one day hope to develop a computer that works similar to the human brain.

Factors that have hampered the more widespread use of PET investigations are the elaborate hardware it needs and the number of expert personnel needed to perform the procedures. Since the isotope labels that are used in the process are short-lived, they must be produced right at the site where they will be used. Their production requires the use of a cyclotron, a massive device that produces isotopes by generating high-velocity collisions between accelerated elementary particles. This process usually requires the presence of a physicist, and, all in all, about three to five highly trained people will have to be involved.

SPECT (Single Photon Emission Computered Tomography)—As with the PET process, the patient inhales, swallows, or is injected with a radioactive substance. A ring-shaped frame supporting a large gamma camera (also called a SPECT camera) rotates around the patient's torso at the area to be studied. As particles are emitted from the radioactive substance and are captured by the camera, their signals are processed and analyzed by the computer (as in the PET process) to form a cross-sectional image of the area of the patient's body under study.

The key difference between PET and SPECT is that the particles emitted in the SPECT process are high-energy photons (photons are tiny packets of electromagnetic energy; see "The Electromagnetic Spectrum") that emerge from the patient's body and strike the cameras, providing only a single reading for each emitted particle. As explained earlier, the radioactive materials used in the PET process emit positrons, which interact with electrons to create two oppositely directed photons, thus providing two readings for each emitted particle. This factor enables PET images to provide about five times more clarity of detail than SPECT.

On the other hand, the SPECT process uses more readily available radioactive emitters—technetium-99m, iodine-123, and thallium-20—which can be obtained without the use of a cyclotron and without the need for the accompanying skilled personnel to operate it. This vastly reduces the cost of the process and makes it much more readily available.

Ultrasound—All of the processes described above use some form of electromagnetic energy as their primary imaging probe. Ultrasonic devices, on the other hand, use mechanical vibrations (i.e., high-frequency sound waves) to obtain their images.

The high-frequency sound waves are generated by having a varying electric voltage activate tiny crystals in the ultrasonic instrument. These crystals then vibrate at the frequencies necessary to produce the required ultrasound.

Because sound waves travel much more slowly than X-rays or radio waves, and require a physical medium such as air, water, or some solid material in which to move, ultrasound imaging has developed along much different lines.

In an ultrasonic examination, the patient will lie prone, while a small instrument is passed over the area of the body to be examined. A gel or lubricant will usually be applied to the skin over that area to provide good transmission for the sound.

The instrument sends the high-frequency sound waves (well above the levels audible to human ears) into the body, where they interact differently with the anatomical structures and tissues that they encounter and reflect back to the instrument. A microphone in the instrument captures these varying "echoes," and the computer uses their recorded values to construct an image of the area under study. Generally, the image is shown on a video screen, and actual movement can be observed.

In addition to tracing structural features in the anatomy, ultrasonic techniques can measure movement, such as blood flowing through the vessels of the heart, very effectively through use of the Doppler principle. This principle established that sound (or light) waves coming from an approaching object are perceived as high frequency, and waves from a receding object as low frequency. (The frequently quoted, familiar example is that of a train whistle, which sounds more and more high pitched as the train comes closer and lower and lower as the train passes off.)

Ultrasound cannot be used to examine the lungs, the brain (with some exceptions), or bones. Some obese individuals cannot be successfully examined by this technique. (Sound waves cannot penetrate pockets of air as in the lungs, and they are distorted by very compacted tissue.)

At the levels used in medical procedures, ultrasonic techniques are believed to be completely harmless to patients. However, as with all medical imaging techniques, constant vigilance is maintained to assure that no unsuspected or long-term effects will develop.

THE FUTURE

For all techniques that use ionizing radiation, medical researchers will continue to seek ways of reducing patient exposure time without impairing effectiveness. CT scans will continue to improve and to provide invaluable insights into medical problems.

If the cost of running PET scans can be brought down (for example, by eliminating the need for an on-site cyclotron), they will assume an even more important role in fundamental research.

The importance of ultrasound imaging is rising, due in large measure to its (to date) fully accepted assurance of harmlessness. One improvement includes the use of color video images from ultrasound scans. These ultrasound images show how rapidly blood is flowing in different areas by using the Doppler flow depth shading (see the discussion of the Doppler effect above), similar to the colors used on Doppler radar-produced weather maps.

Other related imaging techniques are also being expanded upon and developed, including confocal laser scanning microscopy, in which optically thin sections from semi-transparent objects are examined. (It is often used in reference to cancer research.) Magnetic Resonance Spectroscopy gives biochemical information from the patient, using a magnetic resonance scanner. (The principles are similar to an MRI, but MR spectra is analysed). It is used in metabolism studies; in addition, MRS of the brain is used for the study of Alzheimer's, tumors, head injuries, epilepsy, and many other brain-related illnesses. Other research is being conducted to develop electronic image intensifiers, so the patient is exposed to lower doses of X-rays during certain diagnostic techniques. The method uses a lower dose of radiation to obtain an image, then enhances the image with the special intensifiers, giving doctors a better view of the problem area without added radiation.

Still other medical imaging techniques are being used in combination. For example, researchers have combined the information about bone from a CT scan and a tumor from an MRI scan to determine the position of the tumor relative to the bone.

Many experts feel, however, that the star of the imaging field is magnetic resonance imaging. This assessment is true not only because the MRI process is free of ionizing radiation, but also because of the inherently greater information-carrying capacity of its signal pattern.

One major obstacle to its more widespread use is cost. MRI is a very expensive process. One of the key factors in that high cost is the need to use helium-cooled superconducting magnets. Recent developments in high-temperature superconductivity (+) could eventually bring significant reductions in the cost of operating MRI equipment. One such development was a successful brain scan utilizing high-temperature superconducting magnets carried out in an experiment performed at the Hammersmith Hospital in London. This promising result bodes well for the future of MRI.

Other technologies are also helping the progress of medical imaging techniques, especially advances in digital imaging (+), computers, and fiber optics (+) technologies. These developments provide new ways to manipulate and enhance images and allow the doctors to see important details for proper diagnosis and care. New computer architecture is being developed that will allow images from CT and MRI scans to be converted to an internal database format. This

ability allows the doctors to easily transfer the images to other colleagues over a network, and provides a place to store the images until they are needed. New fiber optics (+) lines are making long-distance videoconferencing with remote medical experts (called "telemedicine") a reality, as the pictures from medical imaging techniques are crisper over the fiber optics lines.

FURTHER READING

Books
Foundations of Medical Imaging. Joie P. Jones, Manbir Singh, Z.H. Cho. New York: Wiley Interscience, 1993.
  Reference showing medical imaging as a whole.
Inside Information: Imaging the Human Body. William A. Ewing. New York: Fireside, 1996.
  State-of-the-art imaging techniques showing the inside of the human body.
MRI Physics for Radiologists: A Visual Approach. Alfred L. Horowitz. New York: Springer-Verlag, 1995.
  A detailed look at the physical fundamentals of MRI.
Naked to the Bone: Medical Imaging in the Twentieth Century. Bettyann Holtzmann Kevles. New Jersey: Rutgers University Press, 1996.
  A look at the history of X-rays and the new imaging techniques.

Articles
"A Fantastic Voyage Through the Human Body." *Life*, February 1997, 33-41.
  Imaging techniques show the insides of the body.
"When Memories Lie." S. Richardson. *Discover*, January 1997, 50.
  How PET scans reveal false memories.

Micromachining and Nanotechnology

BASIC DESCRIPTION

The entire concept of micromachining and nanotechnology is based on micron- and nano-sized objects. All manufactured products are made from atoms, and their properties are based on how the atoms are arranged. If scientists can somehow rearrange the atoms in a specific manner, changing the building blocks of nature, they may be able to create anything desired. This is the ultimate goal of such small-scale designs.

Micromachining generally refers to building smaller machines, usually microns, or millionths of a meter, in size. (See "The Metric System.") Nanotechnology is essentially engineering and manufacturing on a scale of nanometers, or billionths of a meter. The great dream of micromachining is to build much smaller electronic or mechanical devices, thus making an overall machine (such as a computer) much smaller; the great dream of nanotechnology is to build machines the size of molecules, which will eventually be able to create any desired end product by rearranging and assembling material on the atomic level.

It is hoped that further developments in micromachining and nanotechnology can be applied to the fields of engineering, physics, electronics, optics, material science, biomedicine, and medicine. At this time, micromachined devices are not too common, and most work in nanotechnology is in the research or development phase. Scientists have great hopes for micromachining and nanotechnology in the future, but progress in both studies has been slow.

Note: Nanotechnology, as it is currently used, encompasses a broad range of concepts. The term is most often used to describe: (1) micromachining, although in a strict sense the term "micro" applies to things that are larger than nanoscales; and (2) the actual equipment that is used for the manipulation, such as the scanning tunneling microscope or the atomic force microscope. In our case, we will use nanotechnology to mean anything on a scale of nanometers.

REPRESENTATIVE APPLICATIONS

Commerce

Micromachining has had its greatest impact in the commercial fields. With the use of advanced lithographic and etching techniques to modify surfaces of materials, machined parts only a fraction of a micrometer in size can be built. One recent promising development was the building of a 10 micron-sized guitar by the Cornell Nanofabrication Facility—the smallest device yet developed and evidence that scientists may eventually develop nano-sized devices for commercial and industrial use.

It is hoped that micromaching and nanomachining will result in micro- and nano-sized small gears, transistors, electronic circuits (+), and other devices that will one day be used in automobiles, microrobotics, telecommunications, spacecraft, and computers. Such small components will allow the machines to be much smaller in the future. For example, microelectronics will continue to shrink the size of computers while adding more processing power than computers that once filled entire rooms.

Research

In the limited sense in which the term "nanotechnology" is being used at the present time, a number of ongoing research applications are achieving some measure of success:

- Simple Atomic Manipulation (for example, such atomic manipulation will help chemists synthesize more complex and useful molecules, and will help material scientists make more useful solids, such as metals)
- Protein and Molecular Engineering and Design (for example, in biotechnology [+], computer-controlled molecular tools smaller than a human blood cell could remove obstructions in the circulatory system or kill cancer cells in the body)

Medicine

One area in which miniaturization, especially in microelectronics, has shown promise is in medicine. Certain types of equipment need to be very small to enter blood vessels; thus, micro- and nano-sized "equipment" would be perfect to monitor various vital patient data. For example, mini-sensors now exist that measure blood pressure in an artery for patients that need such monitoring.

Electronics and Microtechnology

Although current computer chip transistors cannot be seen with the naked eye (they are usually micrometers in size), researchers are now attempting to make transistors even smaller. One type of transistor is about 150 nanometers long and is called a metal-oxide semiconductor field-effect transistor. Smaller devices allow manufacturers to pack more transistors onto computer chips for more memory and faster speeds, with some researchers hoping for memory chips with more and more transistors to create memories in the four gigabits (billion bits) or higher range. (Today's most powerful chips have a capacity of about 16 megabits.)

BACKGROUND

Nobel Prize-winning physicist Richard Feynman challenged the world in 1959 by proclaiming: "The principles of physics, as far as I can see, do not speak against the possibility of maneuvering things atom by atom." His other challenge was to make a working electric motor that would fit into a cube with sides less than four-tenths of a millimeter long. The resulting motor was built by researchers less than a year later, with the motor generating one millionth of a horsepower. But Feynman's request was a powerful catalyst to the world of micromachines and nanotechology, with current research moving into dimensions a thousand times smaller, including manipulations at atomic scales.

Note: Another goal of nanotechnology was proposed and studied by John von Neumann in the 1940s: self-replicating manufacturing systems. These systems would make copies of themselves and manufacture other useful products; this sort of self-replicating manufacturing is one of the long-term objectives of those pursuing nanomanufacturing.

The term "nanotechnology"—a term introduced in 1974 to describe ultrafine machining of matter—is credited to K. Eric Drexler (a PhD in molecular nanotechnology, who wrote the first journal article on molecular nanotechnology), who came up with the idea of nanotechnology while a student at the Massachusetts Institute of Technology in the 1970s. But at that time, the idea was not readily accepted, especially since the technology had yet to catch up with the idea of creating small machines.

By the 1980s, the arrival of (and advances in) such technologies as the scanning tunneling microscope and the atomic force microscope paved the way for better understanding and development in the field. With the advancements in these microscopes, moving individual atoms became a reality in the early 1990s. Even though not a single, true nanotechnological machine has been built to date, scientists have succeeded in manipulating atoms, one of the first steps for developing machines atom by atom. Currently, these machines mostly push atoms around a surface; it is hoped that within the next decade that researchers will be able to piece together the atoms and molecules into useful machines.

THE PROCESS

Micromachining is dominated by production methods that use lithography, in conjunction with chemical or mechanical means, to etch away materials, creating a micromachine or microgear with the leftover material. These methods have been used to create microelectronics and microelectromechanical systems. For example, such devices as tiny accelerometers (a device that measures acceleration or force) that measure drastic changes in automobiles, and activate the inflation of air bags in a car crash, are micromachined.

Several sophisticated tools have been used to develop the techniques of nanomanipulation of atoms. One of the main instruments used in moving atoms on a nanoscale is the scanning force microscopes, which use probes that can position atoms or molecules on a nanoscale (see "Microscopes"). These instruments include the scanning tunneling microscope and the atomic force microscope.

"Top-down" and "bottom-up" are names given to the two approaches used in micromachining and nanotechnology. The top-down approach uses subtractive methods that etch away material to make the components smaller; the main method is called e-beam lithology. For example, micromotors are currently created by etching away silicon, creating gears and other machinery to run the micromotor; nanomotors are expected to be made in the same fashion. The bottom-up method involves moving atoms around to create devices, whether they are mechanical or biological. Most of the examples using the bottom-up method are in the developmental stages, including tailored biomedical products, computing and storage products, and unique materials, such as sensors or controllers, for sensitive aerospace needs.

Not that nanotechology is concerned with only building smaller machines. Scientists also found that by varying the size of the grains of certain materials, they could change the normal characteristics of those materials, such as differences in chemical reactions, or they could cause the materials to be made more or less conductive.

POTENTIAL BENEFITS OF MICROMACHINING AND NANOTECHNOLOGY

If you think about it, humans have actually manipulated atoms for thousands of years: We have made stone knives and tools by casting, chipping, and grinding materials—essentially rearranging the atoms on a large scale. On a smaller scale, we take sand, mix it with certain other materials, and create computer chips. All these examples show the manipulating of atoms on a large scale. As time has passed, we have succeeded in producing better and better materials by this manipulation; and now, micromachining and nanotechnology are taking this manipulation to the atomic level.

Today's micro- and nanomanufacturing methods at the molecular level are very basic. One scientist compared it to moving a child's building blocks with boxing gloves on your hands: It's easy to move them around, but hard to build something with the blocks. The ultimate goal of micromachining and nanotechnology is to essentially rearrange the building blocks of nature to make products cleaner, smaller, stronger, less expensive, lighter, and more precise (thus, hopefully safer), and with those products, make our lives easier and healthier.

For example, manipulation of carbon atoms at the atomic level could lead to a shatterproof diamond; the diamond could be used to create lighter and stronger materials. A Boeing 747 made of this material would weigh one fiftieth of a modern plane, without sacrificing any strength. Changed diamonds could also be used in circuitry and computers, as diamonds work better than silicon at high temperatures. This becomes important to chip manufacturers: As chips get faster and faster, their performance is limited by the amount of heat they can dissipate in the circuitry.

The actual probability that these wonderful products will make it to our homes and stores is high. But it will probably take a few more decades to perfect such materials—especially to determine just how to manipulate and rearrange materials on an atomic scale.

Right now, many of the capabilities currently hoped for are, for the most part, theoretical. The biggest problem is our current technology: We do not have any techniques to approach the level of performance needed to produce such products. For example, a chemical synthesis process can produce 99 percent of the reactants to the desired product. This equals an error rate of 1 in 100—which is ten million times less perfect than what is desired when it comes to nanotechnology.

THE FUTURE

Scientists have great hopes for the future of nanotechnology, in particular in the fields of altering materials on a molecular scale, especially to increase strength, endurance, or resistance to harsh conditions. One of the critical factors in the search for nanotechnology is positional control, or how to position the molecules—in paticular, how to get the right molecular parts in the right places.

Computer simulations have been used to determine some potential applications for nanotechnology. For example, at NASA's Ames Research Center in California, simulations on parallel computers are being used to investigate the properties and design space of molecular gears that would be fashioned from carbon nanotubes (nano-sized tubes created from atoms of carbon), with teeth added via a benzyne reaction known to occur with C-60, a carbon molecule commonly known as a "buckeyball"—thus, the theoretical tubes would be called "buckeytubes." The gear, if it could be built, would be powered by the movement of the nanotube's atoms. Unfortunately, scientists have only been able to simulate this process on the computer, as actual manipulation of the atoms is difficult at this time.

Many scientists believe that micro- and nanotechnologies will increase our knowledge of mechanochemistry, or the knowledge of chemical building blocks, chemical reactions, and molecular aggregates (groupings). For example, one recent study involved the development of nano-test tubes. Carbon molecules were shaped into tiny nanotubes and were then used to carry out chemical reactions at the microscopic level. After opening the ends of the tubes, scientists allowed molten silver nitrate to enter, then heated the tubes with a beam from an electron microscope, resulting in a chemical reaction within the tubes. The technique may be used for microelectronic applications, which may need certain chemical reactions to take place at specific spots in an electronic microcircuit.

Nanomanufacturing plant design is also an area of interest. For example, one company has a goal to build one of the key pieces of molecular nanotechnology—the assembler, or nanomanufacturing plant. This system would manufacture bulk materials or arbitrary structures on the atomic level, getting nearly every atom in the desired place. This would lead to better quality control, and fewer defects, and would produce materials faster and cheaper. The actual mechanism to accomplish manufacturing on this scale will either be mechanochemistry, which is a chemical reaction helped over its normal reaction barriers by mechanical (physical) force; or positional electrochemistry, which overcomes the reaction barriers (or creates a chemical reaction) by careful use of an electric charge.

Eventually, scientists hope that all the work in micro- and nanofabrication will lead to nano-self-replication or self-assembly systems. In these systems, nanomachines make copies of themselves; and after the nanomachines are replicated, they manufacture useful products such as computer and other electrical components that need the precision of nano-machining. In other words, as parts for electronic devices become smaller in the future, nanomachines will be necessary to work on and build such devices. By building the manufacturing system and by producing the final goods by themselves, after the initial human intervention, these nanomachines will save considerable time and money.

In the field of medicine, researchers hope to build bacteria-sized robots that would be injected into the bloodstream. These small robots would travel throughout the bloodstream and would be programmed to rid the body of unwanted organisms. The tiny robots could also be used to eliminate the plaque that lines the arteries and is responsible for heart disease.

The biggest drawback so far to these efforts is their reliance on electron beam lithography to create the smaller features. Even though electron beams are suitable for research, the method is much too slow for mass production. In the future, researchers are looking toward X-ray lithography as a more cost-effective method as, at this point, it is faster than e-beam lithography, and thus appears to be a better alternative for mass production.

Despite the progress so far, there are still physical, chemical, and practical barriers to developing the nano-manufacturing techniques and the products themselves. One of the limitations will occur when devices are so small that physical and quantum effects become significant. For example, will excessive heat or cold have a greater effect on such small devices? With quantum effects, how do the smallest particles react—and how would the manipulation of atoms on a large scale change the physical makeup of developed materials? And would these materials be safe? Thus, few nanofabricated products exist at present, and many of the potential future applications are unknown.

FURTHER READING

Books

Beyond 2000: Micromachines and Nanotechnology: The Amazing New World of the Ultrasmall. D. Darling. New York: Dillon Press, 1995.

   An elementary look at the new micro- and nanotechnologies.

Micromachines: A New Era in Mechanical Engineering. I. Fujimasa. New York: Oxford University Press, 1997.

   The history and applications of nanotechnology.

Nano: The Emerging Science of Nanotechnology. E. Regis (ed.). Boston, MA: Little Brown & Co., 1996.

   A general guide to nanotechnology and how it may be used.

Nanosystems: Molecular Machinery, Manufacturing and Computation. K. E. Drexler. New York: Wiley Interscience, 1992.

   A more technical book on nanotechnology and applications.

Articles

"Making Something out of Nothing." A. Rodgers. *Newsweek,* March 31, 1997, 14.

   The latest in nanotechnology.

"Throwing—or Molding—A Curve into Nanofabrication." R. F. Service. *Science,* July 19, 1996, 312.

   George Whitesides' work in nanotechnology.

Networking

BASIC DESCRIPTION

Networking is the linking together of individuals, organizations, or devices, sharing some form of common purpose, in a connected system of computers and terminals. LANs (local area networks) join people and devices within a limited area, sometimes only a single building. National and international networks such as the Internet provide communications links across the country and the world.

Because networks join such devices as fax machines, printers, and computer terminals, it's possible to transmit pictures, charts, diagrams, and engineering drawings as well as simple text. A vast communications medium known as "e-mail" (or electronic mail) allows the rapid exchange of "paperless" letters from one distant point to another. With special computer software and hardware, it has also become possible to hold live meetings among individuals dispersed around the country and around the world. Networks make use of every available form of signal transmission device, including conventional telephone lines, radio, satellite relays, and fiber optics (+) links.

Because large networks encompass a variety of hardware devices and software systems, maintaining an uninterrupted flow of signals through them represents a formidable engineering task. But as technologies improve, such as additional fiber optic lines for telephones, the ability to send clear signals, and more of them at once, will also improve.

REPRESENTATIVE APPLICATIONS

Commerce

Many companies with dispersed branches are now linked by intra-organizational networks. Electronic mail, accounting reports, in-house orders, and live conferences are handled through such networks.

Industry

Departments in many large factories are linked by LANs (local area networks) for checking inventories, machine availability, and absentee reports. Components in plant security systems are linked with other components and with control points via networks.

Government

Regulatory bureaus were among the early users of computer networks. In many instances, rulings on specific questions were "broadcast" to the branches, via the network, to assure timely delivery.

Media

News articles generated at one point can be published simultaneously at distant locations, using text-setting software and network distribution.

Medicine

Many health professionals are now linked by national networks to other experts within their own specialty. Some hospitals are linked to each other and to health information services.

Military

Sophisticated sensors (+) are linked with large networks of computers in defense and surveillance installations.

Science

The computer network concept is particularly important to scientists, because it enables them to communicate and to rapidly and easily exchange timely or large masses of data and graphics with widely dispersed colleagues who share their research interests. In addition, it presents them with the opportunity to utilize specialized (and very expensive) equipment such as supercomputers (+), which may be at distant locations. Experimental data can be transmitted from one laboratory computer to another (next door or across the country) for analysis and study.

General

Networks among hobbyists, writers, professionals of all kinds, religious organizations, athletic groups, and academic scholars are only a few examples of this rapidly expanding high-tech phenomenon. So far, the Internet is the largest network of all. All one needs is a relatively current computer and a modem (above 9600 baud) to access the hub of this activity.

BACKGROUND

The concept of networks certainly predates the computer age. The telephone system was (and is) a massive, multi-terminal network. In its primary role, it is a relatively homogeneous collection of hardware, dedicated to voice transmission. However, at the present time it is also the major carrier of nonvoice data in the form of computer messages and fax machine memos. It remains the key component in all of the highly sophisticated linkups that are now forming all around the world.

The linking together of people through computers brings a level of depth, versatility, and continuity to communication among physically separated individuals that conventional telephone use cannot hope to match. In some respects, network conferencing seems to create a measure of focus around the issue at hand that face-to-face contact around the table does not always achieve. In addition to ongoing dialogue, charts, photographs, statistical tables, and drawings can be extracted from computer memory and displayed on the CRT screens of all participants. The full power of this still-developing medium has not yet been fully assessed.

THE PROCESS

Local Area Networks (LAN)

In a local area network (LAN), signals may travel entirely by cable within a building or complex owned completely by a single organization. A LAN would generally be operating within an area of a few miles or less and might well involve only the use of equipment and lines that are entirely under the control of the same organization. This control of facilities vastly simplifies the process requirements, because the system does not have to interact with outside utilities and systems, which may have their own complex rules and equipment specifications.

LANs can range in size from a few computers hooked together to share other devices, such as a printer, or hundreds of computers (called nodes in the LAN system) attached to many outside devices, such as printers or plotters. These larger networks are run by "servers" that have extra-large hard drives to help control the LAN system traffic. Another networking method to access connections and data at higher speeds is the Asynchronous Transfer Mode (ATM), which is based on an asynchronous communication between computers. This "start-stop" transmission, unlike a LAN's continuous transmission, allows for fewer bottlenecks for the multitude of users accessing the system.

Wide Area Networks (WAN)

In a large network linking a group of people or organizations dispersed across the nation (or across the world), the problems of equipment and channels become very significant. WANs are built to handle the larger links that cut across such long distances and to solve hardware and software complications that result from hooking up disparate users with different systems.

The task of the architects of large-scale computer network systems is to create a highly reliable array of two-way information paths through a wide assortment of communications devices. These devices include hardware modules from many different manufacturers, software packages from diverse sources, and the in-place facilities of public and private utilities. Redundancy is essential: if one device or channel fails, another must be ready to take its place immediately.

THE INTERNET

Millions of users are "on" the Internet, an electronic link between computers that spans the world. With the Internet, a computer used in the United States can communicate with a computer in England, Russia, or Australia. The Internet is the best example of the interlinking of many WAN systems: Computer information from all around the world, using many different protocols (how a computer communicates with other computers), is processed through worldwide gateways. These gateways allow each protocol to be translated, enabling text, sound, and graphics to be relayed from the source to the connection site.

The Internet started as a project for the United States Department of Defense, which developed a network of computers called ARPnet (Advanced Research Projects) for military and government purposes. Soon after, universities and other institutions began their own networks, and eventually merged these networks with ARPnet to form the Internet.

The major reason for the exponential use of the Internet in recent years is the ease of use. Most computers have communications software that allows easy access to the Internet and modems (which stands for modulating and demodulating) that transmit the data over the phone lines. To connect to the Internet, the user must also have computer access to an Internet Provider (IPs), which usually provides a local-access phone number to the Internet. In addition, the information available over the Internet is not only from the public sector, but also linked to servers of commercial concerns (for example, making it easy to order products online), industry (providing information about their processes for those seeking such technologies), and institutions (allowing users to seek out general and specialty information from universities, organizations, and other sources). One of the most popular connectors within the Internet is the World Wide Web, which provides a place where businesses, public concerns, and individuals can display graphics, audio, and video to enhance the information offered in their textual documents.

The biggest limitation of the Internet is its inability to handle a large number of users at once. Such traffic often creates a bottleneck. For example, when the Mars Pathfinder landed on Mars in July 1997, millions of people tried to access the National Aeronautics and Space Administration's (NASA) server for images and information on the red planet. Because the huge volume of connections overloaded the system, it was difficult to connect to the NASA server or, if connected, to obtain information at the usual downloading speeds.

THE POWER OF NETWORKING

To envision the complexity and power of the WAN and LAN networking concept, let us consider a hypothetical network linking health organizations in the United States and Europe.

Let us say that a high-level American official is stricken with a heart attack while traveling in Europe. At the request of the European doctors, an American hospital where the official has been recently treated sends his records to Europe.

The patient's medical history and a digitized (see "Digital Image Processing") chest X-ray are entered into a computer terminal at the hospital. The terminal transmits this data to the hospital's mainframe computer, which forwards it by telephone line to a ground-to-satellite transmitter located a few hundred miles away. The ground transmitter beams the data up to an orbiting communications satellite (+) via radio; the satellite bounces it down to a ground receiver in the country where the patient is hospitalized. The data then proceeds, via a fiber optics (+) link, to the mainframe computer at the hospital where the patient is currently being treated.

A few minutes after it is sent from the American hospital, the information is being studied by the European doctors.

It turns out that the patient's medical situation is very complex, and that it will be necessary to consult with additional specialists on both sides of the Atlantic. These specialists are located at some distance from the two hospitals currently in contact, but all are members of the same network.

Within a few minutes, the new participants are taking part in a general conference, via keyboard and terminal. They are able to review the medical data, including X-rays and charts, on the computer display screens. All participants can see this data, as well as the comments and recommendations of their colleagues, because they are also printed on the display screen. If those who have just joined the conference wish to ask a question or make a comment, they simply enter it on the keyboard and it appears a few seconds later on the display screens of all other participants.

At one or both of the hospitals involved, participants may also access online databases (+) within local networks for any current published information that might be relevant to the patient's condition. Or they may connect to the Internet and use database-driven search engines to find information or papers important to the patient's condition.

We are glossing over language problems in this little example, except to note that much international technical discussion is now conducted in English. In any case, because the exchanges described here would take place in printed or graphic form on the display screen, problems with accents would be avoided.

SECURITY

The builders of computer networks are invariably faced with two sharply conflicting objectives: first, the need to make access to, and utilization of, the network as simple and friendly as possible for authorized users; second, the need to keep unauthorized users, with unpredictable motives, out of the system. Both objectives are formidable in themselves; when they must be attained in parallel with each other, the designer is faced with a nearly impossible assignment.

A "friendly" system is one that is reliable and reasonably easy to use. Most people are willing to dial the 11 digits required to make a conventional long-distance telephone call. This system is still "friendly." But if it is necessary to dial an additional 15 or 20 digits to provide identification or billing information (as some systems require), the user begins to think about alternatives. (Entering one incorrect digit can mean starting the whole process all over again.)

If, in a networking system, users are required to enter much more than a password and a brief ID number, they may start to avoid the system. (Although with most Internet connection softwares, ID and password, once typed in, are automatically sent to the server each time the user logs on.) If they must pass their text through an encoding machine or enter lengthy identification numbers, they begin to have second thoughts. They may either figure out how to "beat the system" or turn to other means of communication or rival information providers.

Those charged with network security, therefore, must find the middle path between the needs of the user and the integrity of the system.

The simple fact is that any computer that links to a telephone system (networking computers must do so) cannot be considered completely secure no matter what measures are taken. The problem is that the system is always there, like a mountain, and the intruder doesn't have to climb it today or tomorrow or at any specified time. It's not a moving target; it's a stationary target, with no time limits set for its would-be violators. Sooner or later the most sophisticated cipher or hardware padlock will be broken if the value of the sought-after information is high enough.

But we have learned to live with telephones; we know that cellular telephones are insecure, that metal cable phones are a little bit better, and that fiber optics phones are better yet; but none are secure. In spite of this, we can, with discretion, still conduct a tremendous amount of business and scientific exchange on them. But it would be foolhardy to expose truly vital information over the telephone. It would be equally foolhardy to expose any truly vital information over a computer network.

Protecting the confidentiality of information is only one of the computer network's problems. Its other big problem is protecting itself against sabotage. The very power and speed of the network makes it possible to wound it badly if an intruder can figure out how to use all that speed and power for mischief. Unfortunately, this damage can sometimes be accomplished by inserting a self-replicating "infection" program into the system. This type of program may carry instructions to spread itself over the entire network and either wreak real havoc by wiping out or distorting data, or cause less damage by merely sending "humorous" messages to the CRT screen. Such intruder programs have been aptly named "viruses," and they remain a serious problem to network operators and users.

So the networks have to deal not only with people who are trying to steal information, but also with individuals who simply want to disturb or damage the system.

It's difficult for most of us to understand why an individual who is intelligent enough to gain illicit entry to an important computer network would want to play the role of vandal. However, it's important to note that many of these intruders first get into the system by simply stealing passwords: so their intelligence is tainted with a touch of larceny. In any case, systems designers must build in the best possible safeguards against this kind of intrusion and devise methods of containing it when those safeguards are defeated.

THE FUTURE

Networks are coming into existence and spreading so rapidly that it is difficult to keep track of them.

LAN advances will continue in the future, especially with the advent of better connections and computer hardware to increase bandwidth and speed up data transfer. Because such systems are so data intensive, LAN systems will need work in the following areas: Better scheduled backup of data needs to be addressed, as large amounts of information can be lost rapidly as the amount of data grows faster than storage space available; new ways to maximize existing storage space, without sacrificing time or money, will need to be developed; and better archiving techniques, or ways of organizing files so data can be kept for long periods, must be found.

It is believed that the Internet will become more important to the consumer in the future, with several researchers and pundits pointing to the integration of the Internet, telephone, and television into the so-called information highway. Such access would allow users to choose what type of information they desire, and then view the results at their leisure. Because of the ability to track what is downloaded in such a system, companies will be able to better target their markets.

As to the overall future of the Internet system, it is difficult to predict. So far, it shows little sign of a downswing in activity, especially commercially. For example, in April 1996, the number of commercial sites in the World Wide Web increased about 60 percent from the beginning of that year. Online sales reached about 3 billion dollars in 1996; and they are expected to reach over 48 billion dollars by 1998.

But the biggest concern about the Internet—one on which no one agrees on how to solve—is the ability of the system to handle the millions of more users who will be connecting to the system. Some predict that the Internet will eventually crash in a cascade-like manner; other say that with the number of new technologies being developed and with better fiber optic (+) phone lines, the Internet will continue to grow and survive.

Computer networking opens a new window onto the world of information exchange. Its utilization is growing, and its long-term impact will be profound.

FURTHER READING

Books

Absolute Beginner's Guide to Networking. Mark Gibbs, Todd Brown. New York: Sams, 1994.
>A graphical tour of networking.

The Complete Idiot's Guide to Networking. Daniel T. Bobola. New York: Alpha Books, 1995.
>An easy-to-read tutorial that teaches users how to use the LAN technology.

High-Speed Networking Technology: An Introductory Survey. Harry J.R. Dutton, Peter Lenhard. New York: Prentice Hall, 1995.
>A broad overview of the emerging technology of very-high-speed communications.

The Internet Book: Everything You Need to Know about Computer Networking and How the Internet Works. Douglas E. Comer. New York: Prentice Hall Computer Books, 1994.
>Presents computer networking and the Internet from a non-technical perspective.

Network Security: Private Communication in a Public World. Charles Kaufman, Radia Perlman, Mike Speciner. New York: Prentice Hall, 1995.
>An account of the latest advances in computer network security protocols.

Articles

"Getting Together Bit by Bit." Joseph Palca. *Science,* April 13, 1990, 160-62.
>A news update on the rapidly expanding field of networking.

"Managing the Well-Tempered LAN." William Stallings. *Byte,* April 1990, 275-83.
>Local area networks.

"Surfing for Science." C. O'Malley. *Popular Science,* September 1996, 54-60.
>The Internet and science.

"Talking on the Computer Redefines Human Contact." John Markoff. *New York Times,* May 13, 1990, 1.
>Sociological impact of networks.

"Zooming Down the I-Way." A. Reinhardt. *Business Week,* April 7, 1997, 76-80.
>A discussion of the Internet.

Online Databases

BASIC DESCRIPTION

An online database is a collection of useful information maintained in computer memory and organized for convenient search-access for the benefit of dispersed users. Users access the information base through remote terminals connected (usually by telephone line) to the computer. Specialized online databases have become vital sources of information in many fields, including health, science, and commerce.

REPRESENTATIVE APPLICATIONS

General

Very few enterprises and organizations are still too small to have a computerized database. If you call a mail order house that you have used before, or even just your dentist, you'll usually find that they have your address and telephone number (and other relevant information about you) at their fingertips. Whomever you are talking to is looking at a terminal screen that has just plucked the information from a database.

Huge governmental bureaus like Social Security and the IRS have incomprehensibly vast databases. Credit agencies, airline reservation services, law enforcement offices, and state motor vehicle bureaus all have massive online databases.

In large organizations, essential information about employees, products, and customers is kept on one or another of the organization's computerized databases.

Specialized Databases

Specialized databases, covering selected domains of information, are maintained by both commercial and nonprofit organizations. They cover a wide range of subjects including medicine, law, chemistry, biotechnology, engineering, physics, news, business, and a host of other topics.

Commercial databases earn income by charging for "connect time," i.e., the interval between the time the client starts searching in the database and the time that the client checks out. The nonprofit databases may be free to authorized users, except for telephone costs. They are usually supported by professional societies or government agencies.

BACKGROUND

In the early 1960s, as commercial firms began to acquire smaller computers, the concept of "time-sharing" began to develop. Time-sharing allowed terminal and small computer users to hook into a mainframe (a very large, powerful computer) and share its capabilities with other users. People began to think of the mainframe as a "utility," i.e., a general service facility. Among the services offered by such "utilities" was "information." It soon became apparent that well-organized, readily accessible information was a valuable commodity that people were willing to pay for, and commercial online databases began to appear. For the first decade or so (the 1960s), the use of these databases was largely confined to universities and large business organizations. Their use is now widespread, and many small organizations and individuals use them.

Internal Databases

Computerized online databases that are maintained within business organizations, universities, and governmental agencies for internal use differ in two significant respects from the oak and steel filing cabinets that they have replaced:

1. The information that they contain is instantly available to anyone, anywhere who has a connected terminal and an authorized password.
2. Everyone who uses the system sees the same information.

In the days of file folders and steel drawers, this uniformity was not necessarily the case. People, out of necessity, maintained their own files, because it was not convenient for them to use the files of other departments (because of physical distance, differences in file organization, etc.).

Multiple files, covering the same material within the same organization, were commonplace. Such duplicate files, which were meant to contain identical information, rarely did so in practice. Changes took longer to get to one file than to another and, in some cases, never even reached all the files. People made decisions and acted on inaccurate data. Computerized databases have all but eliminated this problem. An update or change to the file is available to all users immediately.

The use of intra-organizational computerized databases does not necessarily mean that more information is being kept than in the past. It does mean that the information is better organized and more accessible. Except in terms of efficiency, such databases are not a radical break with the past.

To enter either an intra-organizational or a commercial database, a user needs a personal computer or a terminal and a modem (a device used to convert computer signals to telephone signals). The user "dials up" the database, either through the computer or by using an ordinary telephone keypad. Once connected to the system, the user enters the database by keying in, on the computer keyboard, a password and a charge number.

External Databases

Commercial and nonprofit computerized databases represent a revolutionary new resource for scholars, scientists, business executives, and often, the public in general. They are a true product of the age of high-tech. Their growth is a measure of the increasing recognition of information as a vital commodity in the contemporary world. The most important development in external databases has been the Internet, which offers much easier access. Instead of calling the actual commercial or nonprofit group's telephone number to log on to their database, the Internet offers easier, and often faster, links to such databases.

The commercial or nonprofit group can either choose to keep their databases only available on their computers, and allow certain external users to access the database (usually through the use of a password); or they can choose to become a "server," in which the database is maintained at the group's location and is made accessible to anyone on the Internet through a special page, called a home page, on the World Wide Web. (See "Networking.")

Commercial and nonprofit database-producing organizations undertake two fundamental tasks:

1. They gather, organize, store, and keep current, under a single structure, all of the essential information pertaining to a specified domain of knowledge. Databases that provide excerpts or summaries of publications require the services of expert indexers/abstracters and must set consistent rules on how their data will be stored and indexed for users. The tendency to provide full-text files once presented databases with a very labor-intensive task, but recent advances in OCR (+) equipment, commonly called scanners, which reads text into storage by optical scanning, has lessened the burden of data entry into the database. Improved storage capabilities have also freed up more space for these full-text databases.

   The database files may not necessarily be at the same physical location; in fact, some are widely scattered. The user is not necessarily aware of this dispersion because all files are accessed through the same interface.

   In some fields, particularly in science, medicine, and technology, specialized databases represent a landmark event in the advance of human knowledge. It means that individuals working within the same discipline, even though separated by thousands of miles, can consult with each other, confident in the knowledge that all have had access to the latest reports of advances and discoveries in their field. In fact, with the Internet, documentation that would have been only shared between colleagues can now be shared online with members of the public if desired.

2. The database-producing organization also undertakes to provide the necessary search mechanisms to enable the user to quickly locate the information sought within the vast reach of the database.

   If the user has complete identifying information about the data that is being sought (e.g., the name of the author and the title of a published paper), the task of finding it in the database is a simple one. But if the information is fragmentary, or the user only wants to "browse" the database within a selected area of interest, sophisticated search tools have to be provided by the system.

   If the user, for example, wants a listing of all articles that have the words "tomography" and "emission" in their titles, the system will quickly scan its entire database for these keywords (which might consist of the tables of contents for hundreds of publications over a period of years) and flash up the response on the computer screen within seconds.

Until fairly recently, most databases have provided only abstracts or summaries of articles, rather than the entire text of such material. If the user wanted the full article, it was necessary to enter an online request; a "hard copy" would then be provided by mail or fax. This is beginning to change. Database services have begun to store the entire texts of certain key publications such as the *New York Times*, the *Wall Street Journal*, and the *Washington Post*. The onset of optical data storage (+), with its far greater storage capacity, will accelerate this trend toward full text retention.

On the Internet, searching a commercial or nonprofit database is similar to the process described earlier. After the user enters the organization's page on the World Wide Web, the user can easily browse the organization's specific database, such as the NASA press release archives, by selecting text that will lead from one page to another. Some databases also provide a search area, in which keywords can be entered, with the list of "hits" (places to check) displayed on the computer monitor in seconds.

However, even the most resourceful scientist can find the search task daunting if the information sought is particularly difficult to define in terms that the system can understand or if the most logical search terms result in too many hits. As a result, new professions, those of "online searcher" and "online Internet researcher," have grown up around databases and the Internet. Professional searchers (particularly within research libraries) have developed a high level of skill and ingenuity in "tweaking" the computer's search mechanisms to get the information. A scientist, writer, or anyone else searching for information can turn the problem over to a "searcher," who, more often than not, will solve it.

Although search mechanisms in databases are growing increasingly sophisticated, the chances of finding what you need, even with only the scantiest of clues, are getting better and better, even if you decide to do the searching on your own. In fact, in many cases, the hardest task is to limit the amount of information found to workable levels.

THE FUTURE

Online databases store vast bodies of accurate, timely information within a single (possibly dispersed) structure, one that is accessible to clients in remote locations through a single interface. They are having a profound effect on the way the world conducts research and carries out everyday activities.

The rapidly maturing technology of optical storage (+) will further enhance the importance of databases. At one time, databases using magnetic storage systems provided only bibliographic references to their indexed material. Great as the capacity of such systems has been, database services would have been badly stretched to do more than this. Optical storage, however, has created such an enormous increase in storage capacity that things undreamed of in the past will become possible, especially with the newer types of disk technology; for example, the digital video discs (DVDs), which have much greater storage capacities than conventional CDs. (See "Optical Data Storage.") As alluded to earlier, the problem in the future will become much more a matter of information management than of information storage.

As databases grow in size, improvements in search mechanisms will be vital. Database experts have already coined a term to describe certain types of search software. They are calling them "knowbots," i.e., a union of the words knowledge and robots. Currently, on the Internet, a knowbot, called a search engine, can, with a very skimpy information request, scan the multitude of databases until it has assembled the necessary data. It will turn the vague request into a specific selection of information to help the user better focus his or her informational needs.

One persistent problem with any database is "hacking," in which a person familiar with computer code easily breaks into certain supposedly secure databases. (See "Networking.") As more people gain access to computers and modems, security may become a major concern in the future of databases.

The growth of databases, optical storage, and sophisticated retrieval techniques will expand the role of conventional libraries and thrust additional responsibilities on them. Research and university libraries are already providing online terminals, instruction, and expert search assistance. Many card catalog systems have already been converted into online databases. This kind of service is spreading to public libraries and will require much innovative thinking and organizational restructuring as libraries expand to include not only books and periodicals within their area of responsibility, but also the growing array of high-tech information tools.

FURTHER READING

Books

Evolution of the High Performance Database. Informix Software. New York: Prentice Hall, 1996.

> Industry technologists walk the reader through today's most revolutionary database developments.

Finding Images Online: Online User's Guide to Image Searching in Cyberspace. Paula Berinstein, Susan Feldman. New York: Online, 1996.

> A very comprehensive guide to finding all kinds of images online.

The National Electronic Library: A Guide to the Future for Library Managers. Gary M. Pitkin (ed.). New York: Greenwood Publishing Group, 1996.

> Examines the impact of the electronic environment on the traditional library.

The Online 100: Online Magazine's Field Guide to the 100 Most Important Online Databases. Mick O'Leary. New York: Online, 1996.

> A guide to the most important research databases on the Internet.

Articles

"Information Retrieval in Digital Libraries." B.R. Schatz. *Science,* January 17, 1997, 327-34.

> How libraries are now using the Internet for searching.

"A Model Paperless Library." J.W. Verity. *Business Week,* December 23, 1996, 80.

> Databases and a modern library.

Optical Data Storage

BASIC DESCRIPTION

Optical data storage is the process of recording information in such a way that it can be retrieved later through the use of optical sensing devices such as laser (+) readers, or less often now, microfilm enlargement lenses. Optical data are usually stored in inert form, such as physical marks on the surface of a disc or some other medium, and such data can be virtually permanent if properly cared for. This archival quality, plus the ability to remove the optical data storage medium (such as a CD) from the computer for transport, is one of the major advantages of optical storage. But its single greatest advantage is capacity, as it is possible to store far more data optically than can be stored through its chief alternative—magnetic storage.

REPRESENTATIVE APPLICATIONS

Entertainment

Audio compact discs, recorded optically, have proven to be an extremely popular product because of their freedom from background noise and from the deterioration caused by normal use.

Commerce

Since optical discs can provide audio and video playback, as well as simple text display, they can be particularly effective in sales presentations, conferences, and training classes.

Computers

Optical disc readers (CD-ROMs, or Compact Disc-Read Only Memory) are standard on most current personal computers, allowing the computer user to access larger amounts of data, such as an entire encyclopedia on a single disc. Some newer desktop computers now use an optical disc for their primary storage device. The far greater capacity of optical storage

technology will eventually lead to much wider utilization in the computer field.

Government

Agencies, such as the Social Security Administration, which store enormous quantities of archival data are becoming major users of optical storage.

Industry

The storage of photographs, engineering drawings, and documents requires enormous chunks of computer memory. The huge capacity of optical storage systems makes them ideal for this type of application.

Libraries

Libraries and their databases are prime candidates for optical storage, as optical media have many hundred times the capacity of similar-sized magnetic media. They are immune to magnetic and electrical disturbances and are much less likely to be damaged in the playback process.

Note: From 1990 to 1995, the Library of Congress (LC) conducted the American Memory Pilot program, a way of discovering how to digitally store its vast collection of books, photographs, and drawings on optical discs, all toward the goal of making the LC a universally available library. The program spawned the National Digital Library program, in which photos, prints, documents, sounds, and motion pictures are made available to the public, for example, the collections of California folk music or the New Deal Stage of United States history. This project presents an outstanding opportunity for library patrons and those with access to the Internet to view rare material without any possibility of damage to the originals.

Medicine

For certain patients with unusual illnesses, carrying the optical records of their medical background might be helpful in case of an emergency. In the hospital archives, such information would use up far less space than the paper records currently maintained and would be much easier to retrieve.

The number of new ways of using optical storage devices continues to grow. For example, a tiny (one pound) hand-held CD-ROM reader is now available. Because of its size, it doesn't read standard CD-ROM discs, but uses instead a slightly smaller version. Its disc is capable of storing up to 100,000 pages of data; and it retrieves and displays 10 lines of 15 characters at one time. Clearly, for the traveling scientist or scholar, and for business people who carry catalogs or parts lists, this type of optical data application can prove to be invaluable.

BACKGROUND

Computers generally have three storage systems: primary, secondary, and tertiary. Primary storage is what we think of as "built-in memory" (it's always there when we turn the computer on). It's made up of ROM (read only memory), which contains permanent, unchanging program instructions, and RAM (random access memory), which can hold temporary programs and data. Secondary memory is made up of either removable disks (floppy disks), or a hard disk (not easily removable). Tertiary memory is "offline" memory and may be tape, disks, or some other medium.

For many years, magnetic tape was widely used as secondary and tertiary memory. Because the tape reader is a slow device (to find a particular file you may have to roll through many feet of tape until you hit the right spot), it soon lost favor as secondary memory. It is still widely used offline, particularly as a "back-up" resource, especially for personal computer users.

Magnetic disks, however, have become the well-established, preferred medium for primary and secondary memory. Since the read-head of the disk system can move directly to the desired spot on the disk (as can be done manually on the old, conventional record players), file access is very rapid.

The supremacy of the magnetic system, however, has definitely been threatened by optical disc technology. There are three important reasons for this:

1. The far greater storage capacity of optical discs clearly represents a significant advantage over magnetic media. Most optical systems use a laser to imprint the data marks on a surface; the laser can "write" much more precisely than can the write-head of a magnetic device. Thus the laser can imprint a far greater number of signals (units of information) within a given area of surface.

2. The superior archival qualities of optical media are another significant advantage. Since optical devices store their data in the form of actual physical marks on a surface (like the printing in a book), their data are not subject to the danger of an accidental wipeout from the presence of a strong magnetic field. This threat is an ever-present danger with magnetic devices because their data are stored in the form of easily erased magnetic patterns. In addition, since, by necessity, the read-write head in the magnetic system must operate very close to the surface of the disk, a physical disturbance could cause the head to actually scrape the surface of the disk and wipe out data. The optical system allows the reading head to operate at a much greater distance.

3. The newer optical storage disc drives are much faster (and becoming even more so) than certain magnetic drives. This speed allows the user to access data more quickly.

In spite of its many advantages, one major factor has kept optical storage out of day-to-day use in personal computers: It could not be erased and written over as could magnetic media. This ability is an indispensable requirement for computer users, who are constantly updating their material. However, this situation is rapidly changing, with the advent of erasable optical discs (and the appropriate drives, called rewritable optical disc drives), and with the advent of optical discs offered as primary memory in some current computers.

TYPES OF OPTICAL STORAGE

Optical systems can be divided into four main categories, with several subcategories:

1. CD-ROMs (Compact Disc-Read Only Memory): These are discs with nonerasable data inscribed on them. They have the same size and appearance as audio compact discs and can store entire encyclopedias on a single disc. Many standard reference volumes, such as the *Oxford English Dictionary* and various encyclopedias, are being issued in CD form. Since there is no need to erase these discs, it should certainly be possible to build into them whatever archival quality is required by the application.

Another form of CD is called the Digital Video Disc (DVD)-ROM. The disc is either single or double layered, and is multi-purpose: It can be used for computers (CD-ROMs), audio (CDs), and video, and it is the first optical storage device to work with all three. It also has a great deal of storage capacity; to compare, the DVD stores 13 times more data than a standard CD-ROM. In addition, future DVDs will be writable with the proper disc drives.

2. WORMs (Write Once, Read Many times): These are discs that can have data inscribed on them once, at which point they become permanent nonerasable records. (When the customer receives them from the manufacturer, they are blank.) In effect, this type of disc allows the user to make in-house CD-ROMs. This enables a commercial firm to make a parts list or a catalog of products in a highly compact form.

3. Erasable/Writable Discs: These are discs that can be inscribed, erased, and written over an unlimited number of times. These discs are intended to replace the magnetic drives as the primary and secondary memory of computers. CD format writable drives (including the DVD) and CD-E rewritable drives are becoming more popular. These drives have read/write capabilities, are much larger capacity than conventional computer drives, and are useful for archiving.

4. Combination Discs: These systems use both magnetic and optical systems. For example, the optical Redundant Array of Independent Disk system (RAID), a combination of Magneto-Optical disk drive and multiple disk drives working in parallel, is currently used for organizations with large data storage needs.

To manufacture erasable optical discs, compromises had to be made, and a certain amount of volatility (which would not be present in a "pure" optical system) is present in these discs. They are actually magneto-optical in principle, using a recording material with a uniform magnetic orientation between two plastic sheets. Data signals are formed by selectively changing the orientation of tiny spots over the surface of the disc. The material used is not sensitive to magnetic fields at ordinary temperatures; however, when the temperature is raised, it will respond to magnetic forces. When imprinting on the surface, the temperature of the targeted spot is heated with a laser, and the magnetic drive head changes the orientation

at that point. The orientation patterns form the signals that carry whatever information is to be imprinted.

To remove text so that an area can be re-used, the laser heats the area, the magnetic head reorients the surface to a uniform level, and it is ready to be written over. This system requires two passes by the head over the disc for erase and rewrite. This extra step means that magneto-optical systems will take at least twice as long as pure magnetic systems to perform the erase-write operation, clearly a very significant limitation.

Library and Information Center Storage

Libraries have used optical storage for many years in the form of microfilm, microfiche, and similar media. Since this material is analog (+) in form, it cannot be stored on the computer without further processing. Microfiche and microfilm retrieval systems have to physically handle stored materials, so they tend to be fairly slow and they have many mechanical components. In addition, the film itself may become scratched and brittle after repeated use. These systems are being replaced rapidly as digital optical records take over.

Virtually all of the material kept by libraries can be digitized with a scanner. (See "Digital Image Processing.") The scanner "reads" the surface of the material to be stored by sweeping a tiny laser beam over the entire area and capturing the reflections with a photosensitive monitor. Numerical values are assigned to the reflections on the basis of their color and density. The computer processes these descriptive numbers and then prints them (through a laser) onto the surface of the storage medium (usually a disc).

This is an "encoded" printing , i.e., we do not see actual numbers, only tiny dots or blank spaces, on the surface of the optical disc. (Actually we won't even be able to see these without a microscope.) A dot represents a one, and a blank represents a zero. Thousands of combinations of these two symbols encode the information taken from the original.

When we are ready to view what is on the disc, our optical reader has a tiny laser beam sweep across its surface, capture the reflections in a light-sensitive device, convert them into electrical impulses, and send them to the computer. The computer "decodes" the impulses and displays the original text or image on a video screen, or prints it out on a printer.

It's interesting to note that a conventional book (the kind that libraries have always carried on their shelves) is in every sense an optical storage unit without the associated "hardware." Our eyes scan the surface of the pages (using reflected light from an ordinary lamp rather than from a laser beam), and our

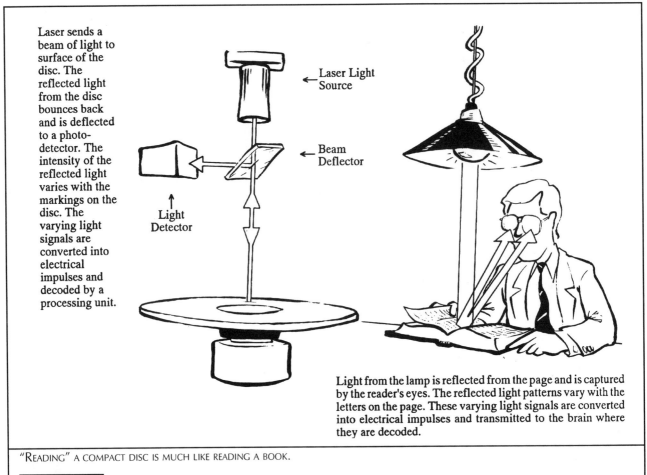

Laser sends a beam of light to surface of the disc. The reflected light from the disc bounces back and is deflected to a photo-detector. The intensity of the reflected light varies with the markings on the disc. The varying light signals are converted into electrical impulses and decoded by a processing unit.

Laser Light Source

Beam Deflector

Light Detector

Light from the lamp is reflected from the page and is captured by the reader's eyes. The reflected light patterns vary with the letters on the page. These varying light signals are converted into electrical impulses and transmitted to the brain where they are decoded.

"READING" A COMPACT DISC IS MUCH LIKE READING A BOOK.

FIGURE 10

brain "decodes" the signals on the page (letters of the alphabet) and determines the meaning of the text. (See figure 10.)

The printing on a page and the dots and blanks on an optical disc are inert signals that won't be disturbed unless we physically attack them or allow them to be exposed to aging chemicals. Magnetic signals, as we have seen, are quite volatile and can be instantly destroyed by a strong magnetic field.

THE FUTURE

A few years ago, magnetic media, in the form of DAT (digital audio tape), may have made something of a "comeback" in the area of offline storage due to its significantly lower cost. But optical storage appears to be winning the day in this area. The tremendous storage density of CD-ROMs, their archival quality, and the ease with which they can be manually handled make them hard to beat.

Initially, because the magneto-optical, erase-rewrite method required a double pass over the disc, it was difficult to see how the method could ever come up to the speeds of magnetic hard discs. However, a process known as the "phase-change" system seems to have solved this problem. In this process the disc surface consists of a thin film of tellurium or selenium. This material exists in both a crystalline and amorphous (noncrystalline) state. The optical reflectivity differs for the two states, and the state can be changed with the heat from a laser. The proper configuration of laser power in the erase-write head of this optical device can perform erasure and rewrite in a single pass. This can be accomplished because the erase-write head consists of two lasers. The more powerful laser melts the material and returns it to, or leaves it in, the amorphous state, no matter what its initial condition was. The less powerful laser converts the spot to crystalline form (or leaves it crystalline) no matter what its initial state was. If the amorphous state represents a zero, the system will write with the high-powered laser to get the zero. To get a one (the crystalline state), the system will use the less powerful laser. (The system deals only with zeros and ones because it is using the binary number [+] system.)

To understand how optical discs will eventually, and perhaps totally, surpass the magnetic hard drives in the future, especially in terms of speed, one should look at the progression of the optical CD-ROM drives just in the past few years: Around 1993, 2X drives dominated; by 1995, 4X (quad speed) drives were rapidly replacing 2X drives; in 1996, 6X and 8X drives took over; with estimates that by 1998 10X and 12X drives will dominate. But such drives are also having competition right now from the Digital Video Disc (DVD)-ROMs and DVD format writable disc drives, which are faster and have higher storage capacities.

New applications for optical data storage are appearing constantly. Although read-only CD-ROM drives represent over 90 percent of all optical disk drive systems offered, writable CD format drives (such as CD-write-once drives) and PD drives (which can use either CD-ROM read-only discs or rewritable discs) continue to grow in availability.

One area attracting a great deal of interest is the use of holography (+) in optical memories. This approach will allow even greater quantities of data to be stored than can now be achieved with conventional optical memories, with some systems reporting up to trillions of bytes on a single storage unit.

These new storage techniques are sure to thrust additional responsibility on the already technologically burdened shoulders of the library profession. But after the initial learning curve and data entry are finished, such storage and easier access to data facilities will only help them in this information-hungry world.

Not all experts are ready to concede that optical media will win the day in computer storage. They point out that access is still slower in optical devices, and that magnetic hard disks are getting faster and developing increased storage capacity.

Although not every information center has the newest in optical data storage systems, prices seem to be decreasing, and speeds and capacity seem to be increasing, making it easier for such technologies to be integrated. It is not yet clear what the continued impact of optical data storage systems will be, but there is no denying that their impact will be profound.

FURTHER READING

Books

The CD-ROM and Optical Disc Recording Systems. E.W. Williams. England: Oxford University Press, 1996.
    A technical overview of CD-ROMs.
CD-ROM Fundamentals. Robert Starrett, Dana J. Parker. New York: Boyd & Fraser Pub. Co., 1993.
    An excellent introduction to the CD-ROM.
The CD-ROM Revolution. Devra Hall, Jerry Borrell. New York: Prima Pub., 1995.
    The impact of CD-ROMs in the coming century.
Welcome to CD-ROM. Tom Benford. New York: Mis Press, 1995 (2nd edition).
    A layperson's guide the world of the CD-ROM.

Articles

"Digitizing the Ancient Near East." N.S. Silberman. *Archaeology*, September/October 1996, 86-88.
    An example of data storage in archaeology.
"Where Libraries Are Leading the Way." Peter H. Lewis. *New York Times*, May 13, 1990, 8F.
    A brief commentary on solutions to the burgeoning problems of information storage.
"The World on a Platter." T. Ferris. *Scientific American*, November 1996, 119-21.
    An update on CD-ROMs.

Robots

BASIC DESCRIPTION

A robot is a device capable of executing a wide range of maneuvers (frequently including the sensitive grasping and releasing of objects), under the direction of a computer. The computer can, through reprogramming or by reacting to feedback from sensors (+), cause the robot to alter its pattern of maneuvers to fit a changed situation or task. True working robots, as opposed to demonstration, entertainment, or science fiction devices, do not look at all like "humanoids." They are strictly functional in appearance and bear a much closer resemblance to machines than they do to human beings or animals.

REPRESENTATIVE APPLICATIONS

Automobile Plants

The most important applications of robots at the present time are in manufacturing plants, particularly those assembling automobiles. Automobile plants, in fact, account for more than half of robot use in the United States. Japanese automakers are even more prolific users of robots than their American counterparts.

Robots typically perform welding, painting, drilling, sanding, and cutting operations on the assembly line. They also move parts from supply to assembly areas and attach modules to the vehicle structure.

Hazardous Environments

Robots can do jobs effectively in environments that would be extremely threatening to human beings: cleaning up radioactive areas, performing tasks within toxic gas atmospheres, extinguishing fires in smoke-filled rooms, disarming bombs, and loading and unloading explosives and toxic chemicals. Robots have also been used in the exploration of sunken vessels at sea.

Laboratories

A rapidly expanding field of robotic activity is in the research laboratory. Small desktop robots, which look like little automated kitchen gadgets, prepare samples, deliver them to the right place at the right time, and mix them with other ingredients in the correct proportions.

Low Light Situations

Many light-sensitive materials (such as conventional photographic films and papers) must be processed in complete or near darkness—a difficult task for human workers. Robots have taken over some of these tasks and will certainly be doing more in the future.

Military

Robot vehicles that can sense targets and (if necessary) take action against them have already been built. Various kinds of surveillance devices have also been produced. Advanced mobile robots, capable of a wide range of independent actions, could play a significant role in future military actions.

Mining

Mining embodies many of the conditions that call for robot involvement. Much of this kind of.work is performed in hazardous environments under low-light conditions. One type of successful robotic mining device operates on a track placed along the mine wall; it mines about 17 to 20 feet of the wall in an eight-hour shift. Even though there are only about 100 of these systems operating in the United States at the present time, they produce about one-third of the coal coming from underground mines.

Printing

In the printing industry, robots have been used to perform miscellaneous tasks around the presses, such as sorting and tying bundles of output material, loading pallets, applying book covers, and delivering paper from storage to the presses.

Security

A growing area of robot involvement is in plant and office security. For example, a very simple device such as a small vehicle with a scanning TV camera, following a predetermined track around a plant or office, can prove to be an extremely effective "guard." More sophisticated devices are on the drawing board, and some are already in production.

Space

Important roles are anticipated for robots in the fields of space exploration (including the recent landing of a rover-robot on the planet Mars, called Sojourner), space manufacturing, and in assorted tasks such as the repair of communications satellites (+).

BACKGROUND

The word "robot" evokes the image of a mechanical device that looks and acts like a human being.

To an even greater extent than with artificial intelligence (+), the concept of robots captures people's imagination. Science fiction books and films and, to some extent, the news media have contributed to the public's combination of excitement and uneasiness about robots by depicting a level of robot sophistication that is not even remotely approached by current technology. It is, however, the understandable goal of science fiction writers and artists to stir the imagination of their readers and viewers, and we do have to remember that many of the "wild" science fiction concepts of a few decades ago have, in fact, become reality.

TYPES OF ROBOTS

It is probably fair to say that, at the present level of development, the more closely a robot emulates the appearance and physical actions of a human being or animal, the less likely that it is capable of doing useful work. Such "humanoids" are likely to be entertainment or demonstration devices. Their "capabilities" are usually little more than gross simulations (they may "talk" via a few canned speeches in computer memory, for example), and they are likely to be following a rigid, limited, and repetitive program.

Factory Robots

On the other hand, the robots that populate factories, laboratories, coal mines, and warehouses around the

A TYPICAL "FACTORY" ROBOT.

FIGURE 11

world, and that routinely perform high-quality, cost-effective work, are completely functional in appearance and do not bear the slightest resemblance to human beings or animals. They are, in fact, likely to consist of a sturdy, rotating vertical shaft embedded in a stationary base, with a single, hinged arm projecting out from the vertical shaft.

There are, of course, many variations on this basic model (some units even have two arms), but most factory robots tend to look something like the illustration in figure 11. In spite of its unimpressive appearance, there is an eye-catching sense of purpose in the way it brings a tool (perhaps a paint sprayer or a welding torch) up to the correct position on the work object (a machine part perhaps, or a partly assembled automobile) and carries out the required operation.

Many thousands of such unglamorous robots are at work in industrialized nations. Many are active 24 hours a day and are producing highly consistent output.

The work that most industrial robots do tends to be narrowly specialized. What differentiates them from conventional machines is that robots can bring their tool, through a distance, to the correct position on the work object and perform a sequence of precise operations within the designated area of the object. The robot works from a computer program: all its actions, and the positions to which it brings its tool, are predetermined by the manufacturing engineers. It is also possible (for some robots) to self-adjust to small changes in the location of the objects that they are working with, such as tools, parts, and work surfaces. This ability requires the use of a sensor (+).

In theory, robots can be reprogrammed to do a substantially different task from the ones they are currently performing. In practice, however, this transformation is a difficult and costly undertaking and doesn't take place very often.

A major pitfall to be avoided by industrial robot designers is that of creating systems that automate inefficiency. Specifically, the designer has to avoid the temptation to build a robot that simply mimics the actions of the human operator it is replacing. Machine capability may make an entirely different (and better) approach possible.

In the Middle Ages, all written text was copied by hand, by people called "scribes." It is possible to speculate that if the technology had existed in that day to assemble robots, the learned centers of the world might have been populated by tireless, error-free automatons, industriously copying the classics and the king's proclamations, while the development of the printing press was set back a few hundred years.

Along these lines it's interesting to note that, in 1773, two French inventors, Pierre and Henri-Louis Jaquet-Droz, a father and son team, built a very human-appearing automaton that was capable of dipping a quill pen into an inkwell and writing a text of up to 40 characters. The automaton was known as the "Scribe." It still exists and still works; it can be seen at the Musèe d'Art et Histoire in Neuchâtel, Switzerland.

In addition to the robotic-arm type of machine, other types of factory robots have proven useful. Robot vehicles transport materials for considerable distances within factory environments. They generally follow a track (usually an invisible one, buried in the floor). While they may self-load and unload, do some sorting, stop when they meet an obstacle, and make deliveries to the right points, such mechanisms are quite low on the scale of robot capability.

Other Types of Robots

Outside of the factory, robots are beginning to perform valuable services in hazardous and inaccessible environments. These activities include bomb disarmament, radioactive cleanup, underwater exploration, fire-fighting, military operations, and handling of toxic substances.

In these applications, the robot is controlled from a distance by a human operator using a radio or fiber optics (+) link. The robot carries a video camera (or cameras) into the hazardous area, thus enabling the human operator to view the remote site with sufficient clarity to be able to guide the robot's actions.

Since the task of specifying those actions on a keyboard or control panel is far from easy, much development work centers around providing the operator with a better means of controlling the robot. One interesting concept is to have the human operator wear an electronic harness and go through the motions necessary to accomplish the particular task. Electronic sensors (+) in the harness convert those motions into signals that are transmitted to the robot, which then duplicate the operator's actions at the work site. Movie makers have also taken advantage of such electronic sensors. For example, to create the lifelike movements of the *Tyrannosaurus rex* in the movie *The Lost World*, an operator strapped into such a harness guided the motions for the giant dinosaur.

THE FUTURE

The use of robots continues to grow as technologies improve. For example, robots with optical character recognition verification, a process developed for machine vision (+), can be used in industry for such tasks as character identification on a line of print. Improvements in camera technology, higher performance microprocessors (+), and new software developments have led to robots that "see" in color, not the usual monochrome. Even the Mars Pathfinder mission used a lander and rover robot developed from advanced computer and digital imaging (+) technologies.

There is growing interest in the use of miniature robots that could perform specific functions in inaccessible places. Small "biobots," usually a combination of insect and robot, are being explored as a way of analyzing how insects move, with the eventual goal of building tiny robots that move with the same grace, agility, and speed as the insect. Several labs are working on a even smaller, multi-legged device with many sensors (+) but only a bare-bones processing capability. The robots' motions, many of them insect-size, are based primarily on reflexes from sensor readings, making the robot appear remarkably lifelike. Advances in such robots is also dependent on micromachining (+) to build the smaller gears and mechanisms needed to run them. It is not quite clear, at present, what practical applications will follow from this research (although some of the smaller robots have been used to detect cracks along gas tanks in Japan), but it has attracted a great deal of interest.

Another type of larger robot currently being worked on are the master-slave robots. A human fitted with sensors moves in certain directions. The motions are measured by sensors and relayed to the robot via a computer, mimicking the human's movements. Although the master-slave robots easily match the human's smoothness and speed, they are totally dependent on the input movements of the human and cannot work or perform on their own.

The foremost goal, however, of many robotics researchers is the creation of autonomous, self-navigating robots that would be intelligent enough to traverse distant areas on their own, with a preprogrammed search program.

Progress with the autonomous robot awaits significant advances in other closely related devices and systems. These systems include high-capability sensors (+), machine vision (+), speech recognition, parallel processing computers (+), advanced software (including expert systems [+]), dramatically improved batteries and/or the development of alternative energy storage devices, and highly reliable and efficient miniature motors developed with the help of new technologies in micromachining (+).

While the factory robot has gained a firm niche in industry, its use is not expanding. The combination of cost (a new robot can cost more than $10,000) and human versatility has tended to confine the robot to those tasks that are dangerous or that humans simply don't like to do.

Nevertheless, there is little doubt that robots will have an increasingly vital role to play in many important areas of human activity. The eventual development of the autonomous robot will certainly bring science fiction fantasies a good deal closer to reality.

FURTHER READING

Books

Android Epistemology. Kenneth M. Ford, Clark Glymour, Patrick J. Hayes (eds.). Cambridge, MA: MIT Press, 1995.
   The theories and possibilities of developing autonomous robots.
The McGraw-Hill Illustrated Encyclopedia of Robotics & Artificial Intelligence. Stan Gibilisco (eds.). New York: McGraw Hill Text, 1994.
   An introductory overview of robotics and artificial intelligence.
Robotics: The Marriage of Computers and Machines. Ellen Thro. New York: Facts on File, 1993.
   Introduction to the science of robotics.

Articles

"A Cricket Robot." B. Webb. *Scientific American*, December 1996, 94-99.
   The science of small robots.
"Robot Revolution." C. Suplee. *National Geographic*, July 1997, 76-95.
   A look at the use and development of robots in different fields.
"Robots on All Twos." B. Moran, L. Van Dam. *Technology Review*, November/December 1996, 99.
   An interesting perspective on two-legged walking robots.

Supercomputers

BASIC DESCRIPTION

Supercomputers are the fastest and most powerful data-processing machines available at any particular time. Initially, the term "supercomputer" only applied to the larger mainframe machines; today, the term also includes parallel processing machines (computers with multiple processors).

Supercomputers are usually described in terms of the number of operations (such as taking the square root of a number) that they can perform within a given interval of time. This figure keeps changing with advancing technology. The newest machines claim close to a trillion operations per second.

Supercomputers are particularly effective where massive quantities of data must be handled, as, for example, in weather forecasting. They are also very useful in simulating physical processes, such as environmental changes caused by industrial activity. This type of program helps scientists and governments to determine what corrective steps may have to be taken. In manufacturing processes, supercomputers help engineers design new products by modeling how various alternative products will perform when they are actually used.

REPRESENTATIVE APPLICATIONS

Defense

Supercomputers have been used to evaluate the design of new weapons systems and to assess the effectiveness of new strategies.

Global Concerns

Supercomputers are playing a major role in the ongoing debate about environmental dangers. Computer models of atmospheric changes, run on supercomputers, are the chief source of information in the current controversies about global warming.

Industry

Supercomputers have already played a major role in the design of aircraft and automobiles; they promise to be an even greater factor in the future. The Ford Taurus was designed with the help of a supercomputer. Most of the American and many foreign car companies now have supercomputers. The major contribution of supercomputers in these areas is their ability to simulate the real world conditions that automobiles and aircraft will encounter in actual use, such as airflow. The supercomputer will never completely replace the wind tunnel in the aircraft industry, but it has become an important supplement to the tunnel.

In the petrochemical industry, supercomputers have become an important resource in the evaluation of seismic data that may lead to the location of oil deposits.

In the construction industry, supercomputers are helping in the analysis of stress factors in structural materials. They have been used in the design of skyscrapers and sports stadiums.

Meteorology

Supercomputers are the most powerful tools ever devised to deal with the huge accumulations of data that flow into weather centers for analysis and forecasting. Increased accuracy and projected forecasts, which have already been achieved, are the result of contributions from the supercomputer.

Research

It is difficult to overstate the importance that the supercomputer has assumed for the scientific and research communities. In astronomy, particle physics, molecular biology, genetic engineering, and cancer and AIDS research, the supercomputer is playing a vital role. A surprising area of supercomputer research has been within certain fields of mathematics. (Some mathematicians have tended to look down on computers as "brute force" number-crunchers, with little theoretical value in their subject.) However, in the investigation of particular irrational numbers and large primes, the power of the new machines has helped to solve previously unassailable problems.

BACKGROUND

The term "supercomputer" was probably first used around 1957 to describe the large machines being produced by Control Data Corporation, Remington Rand, and IBM. They were capable of performing millions of operations per second.

When the Cray-1 was introduced in 1976, it set the design pattern for supercomputers which has dominated the field for many years.

Modern supercomputers achieve their great speeds by combining a number of factors:

- Their components, based in large-scale integrated circuits (chips [+]), are extremely fast. They can switch from one state to another in less than a billionth of a second.
- Their "clock speeds" (the basic rhythm at which the machines operate; it's sort of like the ticking of a clock) are also very fast. This rate reaches more than a hundred million cycles per second. Currently, the fastest desktop computers have clock speeds of about 33 to 200 million cycles per second. (MHz), with several companies promising faster speeds in the near future.
- They handle larger segments of data (called "words") at one time than do conventional computers.
- Supercomputers also use a technique called "pipelining" that permits them to prefetch procedural instructions from memory, so that steps in processing a problem are not delayed when it is time for those steps to be performed.

Although speed and power have long been the bywords of the supercomputer industry, the means of measuring these attributes has increasingly come into dispute. The media frequently report the attainment of a new gigaflops-per-second number by one or another of the manufacturers. ("Giga" is the metric prefix for a billion; "flop" means "floating point operation"; reduced to ordinary terms the expression can be read as billions of operations a second.) Increasingly, however, users are questioning how accurately these numbers translate into effective performance. Some are calling for practical tests more closely related to the tasks that the supercomputer will be called upon to perform.

This intramural debate over the difficulties of even measuring the enormous speeds of these machines points up the much greater problems involved in harnessing their capabilities effectively. For example, if supercomputers are to reach their level of maximum usefulness, they have to be available not only to users in their immediate vicinity, but also to users far away from the sites at which the computers are physically located. This accessibility can be achieved by networking (+). However, networks can transmit data at only a tiny fraction of the speed at which supercomputers can produce it. This mismatch and many other practical bottlenecks are currently being addressed, especially with the advent of fiber optic (+) lines and advances in manipulation of the data stream (or the way a computer sends or receives the data).

Parallel Processing

As stated earlier, parallel processing is often the choice for many researchers who demand the speeds and versatility of a supercomputer. Parallel computers are not only fast and versatile, but they are often more cost effective than the larger mainframe supercomputers.

Most computers work on a "serial" basis, i.e., they perform the steps necessary to solve a problem in a sequential order, more or less as a human being would do—first step, second step, third step, etc., until the answer is obtained. When the problem is a long one, with hundreds or even thousands of steps, this process is very slow. (That is, by the standards of the computer world; by human standards it would still be incredibly fast!) One homely little example that is frequently cited to illustrate this point is that of a restaurant that has one cook and a thousand waiters. Clearly the cook can prepare the meals only in the sequence in which the orders are received from the waiters, and there are long waits for the patrons. However, if a hundred more cooks are hired and meal preparation is divided up into logical segments, i.e., one cook does all the vegetables, one does all the salads, and so on, the waiting time for customers to be served will be substantially reduced.

This is the principle of parallel processing. The big, time-consuming problem is divided up into logical segments that can be tackled simultaneously by many processors. For this reason, most supercomputers are now being built with multiple processors. However, not all problems lend themselves to parallel processing treatment, and even for those that do, the task of preparing them for multi-processor handling is not often a simple one. But these are practical difficulties, not matters of principle, and this technique certainly prevails in a high percentage of supercomputer applications.

THE SUPERCOMPUTER AT WORK

Problem Solving

Supercomputers have broken ground in two extremely important areas. They can solve massive, multi-step, data-laden scientific and mathematical problems that, from a practical point of view, simply could not be handled on a conventional computer or by any other available means. This capability is changing the way scientists and mathematicians think about other conceptual and calculating roadblocks that have impeded progress in their areas. In other words, problems that have been gathering dust because they appeared too diffuse and too complex are now getting a second look. For example, biochemists have long despaired of determining the precise structure of certain large, complex human proteins. Mainframe supercomputers have now been used to solve this and other related problems by being able to analyze the extremely large and complex data sets inherent in human protein analysis.

Simulation

The other area of great strength of the supercomputer is in simulation. In simulation, the researchers create a model of some real (or theoretical) world situation and "run" it on the machine. Thus, instead of testing the mockup of a new aircraft in a wind tunnel, a representation of that experiment can be developed for the computer and "run" in a simulation. At this stage of development there is little danger that the computer will completely replace the tunnel; rather they will work together toward the best solution. But the number of actual physical tests can be greatly reduced, saving a great deal of money.

Many systems are simulated on the supercomputer because they cannot be analyzed directly. For example, they may be hazardous, e.g., the processes taking place within a nuclear reactor; or they may be distant and on a huge scale, such as astronomical events; or they may take place slowly on a barely perceptible day-to-day level, such as a buildup of contaminants in the atmosphere. Simulations in the supercomputer enable us to consider these processes on a reduced scale or within a compressed time frame.

Another important factor in using simulation, rather than physical experiments, in commercial applications is that the computer simulation is much more likely to provide solutions to design problems than will the physical experiment. The interacting components of the experiment can be better understood and visualized on the computer screen than they can (for example) in the wind tunnel. While currents of air sweeping over the wings of an aircraft are invisible in the tunnel, they can (if the simulation is properly prepared) actually be seen on the computer screen. This function eliminates a great deal of uncertainty.

Visualization

While much discussion in the supercomputer community now centers around the concept of parallel processing, many experts feel that "visualization" is an equally important concept. Visualization can be defined as the conversion of the numerical output of the machine into graphic form, which can be more readily grasped and analyzed by human experts. The simplest and most familiar forms of "visualization," to most of us, are simple bar graphs that convert complex statistical data into easily understood, multicolored charts.

Since many supercomputer simulations are models of incredibly complex processes, such as the evolution of stars and the internal operations of complicated, high speed machines, the process of developing lucid and accurate "visualizations" for analysis is no easy task. It is truly a specialty in itself and is drawing on the talents of people outside the computer field, including professional artists.

One such use of supercomputing and visualization is the Computer-Aided Visual Environment (CAVE). The CAVE uses actually processed data, such as the graphical representation of a certain human protein, shown on two perpendicular screens in a darkened room. Using special glasses and a joystick-type device, the researcher in our example can "travel" through the protein's many convoluted strands, examining such particulars as how the proteins hold together.

Another use of visualization and the supercomputer that is still in its infancy is virtual reality. Without leaving the room, a researcher can, with special devices such as stereo goggles and a data glove, interact with a computer-generated world. For example, an automotive engineer can see how the control system of a new car works by taking the virtual counterpart for a ride in the computer-controlled environment.

THE FUTURE

In 1985, the National Science Foundation established five national supercomputer centers at American universities; they included Carnegie Mellon, Cornell, Princeton, the University of California at San Diego, and the University of Illinois. The purpose of this largely federally funded program was to promote the related concepts of networking and supercomputing. By 1997, only three of the institutions were being funded by the federal government, the result of budget cutbacks.

However, there are now many other governmental, private, and academic supercomputing centers. A great deal of interest is growing around the concept of a national high-speed network, in which these centers will play a major role.

It is likely that the role of supercomputers in industry will continue to expand over the next decade. It is also probable that the role of the supercomputer in medical research will become a far more significant factor than it has been so far. The ability of the big machines to simulate processes of any size and on any time scale should provide breakthroughs in the area of chronic diseases with complex causes, such as schizophrenia and diabetes. And if increases in computer technologies continue at their present rate, there may be greater advances down the road.

Another area of interest in supercomputing is the development of optical computers, making the supercomputer less bulky and still capable of carrying large amounts of information. Optical computers rely on light rather than electricity to perform calculations. One idea is to use the massive parallelism of light (light is, simply put, "packets" of photons—and many of these packets can be carried along at once), expanding a laser beam so that it carries millions of pieces of information and then processes the information at the same time, something that is not possible with electronic computers that are based on electrons, not photons.

Still another type of computer that is far into the future is the quantum computer, in which the computer would achieve record speeds by taking advantage of the behavior of elementary particles; it would store more information than a conventional computer by using the potential states of an atom, but an explanation of this would take a course in quantum mechanics! Until physicists determine the true behavior of such particles, quantum computers will be mostly theoretical.

The quest for greater speeds will continue, as will the disputes about how to measure it. In the view of many experts, the three key words in the future of supercomputing are networking (+), parallel processing, and visualization. The concepts embodied in these terms will, in their opinion, provide the framework for further fundamental advances in supercomputing.

An application such as virtual reality (VR) may have to wait before it becomes a major player in many fields. One problem is that there is a substantial gap between the technology available today and the technology needed to bring genuine virtual environments closer to reality. One issue is the comfort of the user. Currently, most VR demonstrations are short, and generally require bulky paraphernalia, such as headgear or uncomfortable datasuits. In addition, few studies have been conducted concerning how people interact with such systems, especially when sound or touch are added (for example, will there be problems with motion sickness?). Using VR for recreation has a long way to go; in the meantime, it is more likely that VR will be used for education, research, emergency situations, manufacturing, and design.

FURTHER READING

Books

Expert Systems: Design and Development. John Durkin. New York: Macmillan Coll. Div., 1994.
> A highly technical reference on expert systems.

Parallel and Distributed Computing Handbook. Albert Y. Zomaya (ed.). New York: McGraw Hill Text, 1996.
> A technical guide to parallel computing.

Supercomputering and the Transformation of Science. William J. Kaufmann, Larry L. Smarr. New York: W.H. Freeman & Co., 1993.
> A state-of-the-art look at the capabilities and scientific contributions of supercomputers.

The Supermen: The Story of Seymour Cray and the Technical Wizards behind the Supercomputer. Charles J. Murray. New York: John Wiley & Sons, 1997.
> The history behind the supercomputer.

Articles

"Probing Cosmic Mysteries by Supercomputer." M. Norman. *Physics Today*, October 1996, 42-48.
> Using supercomputing in astronomy.

"Tackling Turbulence with Supercomputers." P. Moin, J. Kim. *Scientific American*, January 1997, 62-68.
> Applications in supercomputing.

Superconductivity

BASIC DESCRIPTION

Superconductivity is the ability of a material (metals, alloys, organic compounds, or ceramics) to carry an electric current without resistance (and without the associated energy losses), below a certain fixed temperature. This temperature varies with different materials and is referred to as the "critical temperature" for superconductivity. Until very recently, superconductivity could be achieved only at extremely low temperatures (close to absolute zero); at these temperatures a number of materials superconduct, including lead and mercury, which are good superconductors under these extreme conditions. But keeping the materials at such temperatures is impractical and expensive. In the last few years, newly developed compounds have become superconductive at more practical, higher temperatures, under laboratory conditions.

REPRESENTATIVE APPLICATIONS

Most current applications of superconductivity still use low-temperature liquid helium cooling. Since this process is very complex and costly, only a limited number of high-budget systems can afford it. More recently, new ceramic materials have been discovered that become superconducting at about 70 kelvins. The ability to superconduct at this "higher" temperature allow these materials to be cooled with cheaper liquid nitrogen, and thus more applications are being discovered.

Medicine

The powerful magnets used in magnetic resonance imaging (MRI) (+) of the human body are cooled with liquid helium to superconducting levels to prevent heat build-up and energy losses.

Highly sensitive instruments called SQUIDs (superconducting quantum interference devices) are used in medical diagnosis to detect and measure very faint magnetic signals from the human body.

Research

Nuclear magnetic resonance (NMR) devices, which operate on the same basic principles as MRIs, also require superconducting magnets. NMR instruments are used in research laboratories to analyze interactions between radiation and matter.

The medical applications of the SQUID were discussed above. In addition to those applications, the SQUID has many other important scientific uses because it is the most sensitive detector of magnetic fields, and other electrical phenomena, ever developed. It is used, for example, by particle physicists to detect the presence of certain elementary particles, and by astronomers to pick up extremely weak signals from distant galaxies.

Particle accelerators are huge research devices that utilize powerful superconducting magnets to guide high-velocity subatomic particles into collisions with each other. By analyzing the results of these collisions, physicists may discover previously unknown particles or unusual interactions among known particles. This knowledge contributes to our fundamental understanding of the origin and nature of the universe.

BACKGROUND

A conductor is any material that can carry electric current. The most common examples are metal wires and cables. Metals, particularly copper and silver, are good conductors because they have many free or loosely bound electrons in their structures. Electrons are tiny, negatively charged particles usually found in orbit around an atomic nucleus. Electrons can be made to move through the conductor by imposing a force (voltage) on them. As they move, they create an electromagnetic field.

In their passage through the conductor, electrons tend to collide with structural imperfections in the conductor and are scattered, thereby generating heat and losing energy. This effect is known as "resistance." Resistance can sometimes be a useful property, as, for example, when you are trying to heat the filament in a lamp so that it will give off light. In many applications, however, it is an unwanted side effect, wasting energy and creating a heat build-up that has to be cleared from the system.

In 1911, a Netherlands researcher, Heike Kamerlingh Onnes, discovered that when he cooled mercury to a temperature of minus 269 degrees Celsius (minus 452 degrees Fahrenheit), it began to conduct electricity without any resistance. He had succeeded in getting the mercury to this low temperature, which is just a few degrees above absolute zero, by cooling it with liquid helium. (See "Cryogenics.") Absolute zero is the temperature at which virtually all motion ceases within a substance.

Kamerlingh Onnes's dramatic discovery attracted a great deal of scientific interest and eventually won him the Nobel Prize in physics. But the practical difficulties of running any commercial or industrial systems at temperatures so low that they required liquid-helium cooling kept the applications very limited. Only high-budget items like NMR, MRI, SQUIDS, and particle accelerators could afford the expense and complexity of this cooling method.

In the years following Kamerlingh Onnes's discovery, the search for higher temperature superconductors continued, but relatively little progress was made toward this goal.

Other facets of the superconducting phenomenon, however, were coming to light. In 1933, a researcher in Berlin, Walther Meissner, discovered the effect that has come to be named after him. He found that superconductors were completely diamagnetic, i.e., they totally repel external magnetic fields.

If you place a magnet over a superconductor, it will remain suspended in the air, without visible support, because of the "Meissner effect." This effect has served as the basis for what has become a popular demonstration stunt at technical conventions. On a more practical level, it is also the basis for the "maglev" (magnetically levitated) high-speed train systems, which will be discussed briefly later in this section.

In 1958, a theory was offered to explain the underlying process of low-temperature superconductivity.

Developed by three American physicists, John Bardeen, Leon N. Cooper, and John Robert Schrieffer, it is called the BCS theory, after the men who developed it. The theory holds that free electrons, because of certain interactions of force within the superconductor, become bonded into pairs that move together in phase with other such bonded pairs. These pairs, because of a kind of mutual reinforcement, do not scatter when they collide with structural deformities and, consequently, no resistance develops. Although the three researchers were given the Nobel Prize for physics in 1972 for their theory, it has not been completely successful in explaining high-temperature superconductivity, and researchers continue to seek additional clarification.

In 1962, Brian David Josephson, a young British physicist, predicted that when two superconductors are placed close to each other, but separated by an ultra-thin insulator, the imposition of a magnetic field will cause pairs of electrons to "tunnel" through the insulator.

This prediction was subsequently confirmed and led to the development of the SQUID and other important devices such as the Josephson junction. The fundamental importance of the Josephson effect is that magnetic fields so slight that they cannot be detected in any other way will cause measurable "tunneling" in Josephson devices. In medical applications, for example, certain diseases such as hemochromatosis, which causes excessive iron to build up in the blood, can be detected with specialized SQUID devices because of the tiny magnetic fields flowing from the iron. Brian Josephson was awarded the Nobel Prize in 1973.

Meanwhile, the search for higher-temperature superconductors continued in laboratories around the world. However, a high mark of minus 250 degrees Celsius (minus 418 degrees Farenheit) or 23 kelvins, attained by researchers in 1971, seemed to represent a plateau that they couldn't rise above.

Note: In discussing superconductivity, most scientists use the Kelvin scale. The Kelvin scale starts at absolute zero, which can be informally defined as the temperature at which all physical activity ceases and below which no further measurements can be made. The Kelvin scale moves upward from absolute zero, in increments equal to Celsius degrees, so that at 273 degrees Kelvin (or 273.16, to be exact) the freezing point of water is reached. So 273 kelvins equals zero degrees Celsius or 32 degrees Fahrenheit. The Kelvin scale avoids all the minus signs that have to be used in the other systems to describe cold temperatures. Conversion from Celsius to Kelvin is easily accomplished by adding (algebraically) the Celsius reading to 273. All Kelvin readings below 273 are below the freezing point of water, and when they get down to double or single digits they're describing very cold conditions indeed. As a matter of practice, one does not say Kelvin degrees; simply kelvins (not capitalized) or K (capitalized) suffices.

PROGRESS IN SUPERCONDUCTIVITY

In 1986, two researchers working in the IBM research laboratories in Zurich, Karl Müller and Georg Bednorz, announced that they had achieved superconductivity in a ceramic oxide compound at 35 kelvins. Although initially received with caution, this announcement eventually attracted tremendous attention and touched off intense activity in research centers around the world. It led to the common use of a new term: "high-temperature" superconductivity (HTS). Müller and Bednorz shared the Nobel Prize for physics in 1987.

Since the Zurich discovery, many new high-temperature superconducting materials have been reported, including the one mentioned above that uses liquid nitrogen and superconducts at 77 kelvins; this experiment is very significant because liquid nitrogen is much easier to handle than liquid helium, much less expensive, and can absorb much more heat before vaporizing. Still another ceramic conductor, yttrium-barium-copper oxide (YBCO), superconducts at about 95 kelvins, and it also uses liquid nitrogen to cool. Unfortunately, similar to many of the other ceramics being found today for superconducting, YBCO is brittle, difficult to shape, and almost impossible to make into thin wires.

Although real world applications of superconductivity are much closer, many practical problems still need to be overcome. For example, a superconducting cable would have to be capable of carrying adequate current density.

The question of current density is not a trivial one. While temperature receives the most attention in superconductivity discussions, the following additional factors have an important impact on practical applications:

1. If too much current is forced through a conductor, it will cease to superconduct.
2. If an excessively strong magnetic field is present, superconductivity will cease. This point is very significant because, as indicated earlier, many important superconductivity applications involve extremely powerful magnets.

Two other limitations bear on the question of future applications: first, many superconductors are most efficient at below half their critical temperature (the temperature at which superconducting begins), and second, superconductivity cannot be sustained with alternating (AC) current. It is only usable with direct (DC) current. Since most municipal power distribution systems use AC, this dependency on direct current represents one more engineering complication. Many additional problems remain to be overcome before high-temperature superconductivity becomes a realistic alternative to existing systems.

One very significant application would be the large-scale storage of energy. Every industrial society faces the frustrating problem of the uneven consumption of power. In periods of low consumption, generators are underutilized; in high consumption periods they are strained to, and beyond, capacity. In addition, the energy from many intermittent sources such as wind, sun, and tidal forces cannot be accumulated and saved, but must, for all practical purposes, be used as soon as it happens to be available.

Storage batteries do an excellent job where small amounts of power are required, but to store large amounts of energy (the kind of quantities required by a metropolitan area), they would have to be impossibly large.

Other storage methods are under study, such as filling huge tanks with compressed air, which would be stored and used later to drive turbines. Generally speaking, however, the engineering and economic problems inherent in these approaches make them impractical on a widespread basis.

Nevertheless, the prospect of large-scale energy storage through superconductivity is a very attractive concept. Utility companies and governmental agencies are looking at this possibility and have already assigned an acronym to it, SMES, standing for superconducting magnetic energy storage.

A low-temperature SMES unit (one utilizing liquid-helium cooling) has already been designed; it calls for an underground ring with a diameter of a little less than a mile. It could store up to 5,000 megawatt hours of electricity. (This amount would be enough to take care of the power requirements of a city the size of Rochester, New York, for example, for about five hours.) Energy could be fed into such a system in times of low public consumption and drawn out in times of high consumption. Such a system could also serve as an accumulator for intermittent power sources such as wind and sun.

If the SMES unit could be cooled with liquid nitrogen instead of liquid helium, it would become an even more attractive proposition from the engineering and economic points of view.

Another widely discussed future application is the magnetically levitated ("maglev") high-speed train utilizing the Meissner effect. "Maglev" trains would float a few inches above the track, overcoming many of the problems that arise from mechanical contact between the wheels of a conventional train and the track itself; such contact requires that the train's undercarriage be massive and that mechanical tolerances be extremely tight. An inevitable result of this mechanical contact is to limit the potential speed of the train. By eliminating this problem, the "maglev" train travels at very high rates of speed. (See figure 12.)

The Japanese have already built such a system, and this type of transportation represents a real advance for their country, where fast commuter trains are already a fact of life. The Bullet Train, as it is called, travels up to 300 miles per hour. Japanese and German researchers (who are also working on a maglev train) are hoping to break the 500 miles-per-hour barrier in the near future.

In the United States, on the other hand, a great deal of inter-city commuting is handled by personal automobile and small commercial planes. Although the maglev concept is being aggressively promoted by some groups, others point out that converting to rail, where small planes are now providing adequate service, would require large capital outlays, possibly litigious "right-of-way" actions, and would involve changes in people's deeply ingrained habits. They argue that maglev trains might well be a marginal proposition in the United States.

THE FUTURE

Although the real dream is to find superconductors that operate at room temperature, the development of the recent "high-temperature" superconducting materials has brought many new potential applications to the forefront:

- More efficient power transmission
- Large-scale energy storage

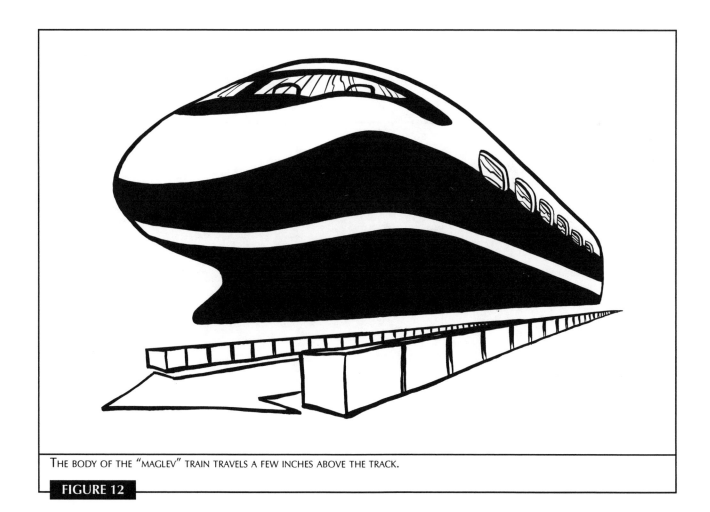

THE BODY OF THE "MAGLEV" TRAIN TRAVELS A FEW INCHES ABOVE THE TRACK.

FIGURE 12

- Magnetically levitated (Maglev) high-speed trains (already a reality in Japan)
- Faster and more efficient computers
- More compact and efficient electronics and motors

Note: Computers and motors would both benefit by operating at lower temperatures (eliminating much heat removal) and would use electric current more efficiently.

In getting high-temperature superconductivity out of the laboratories and into industry and field use, researchers must develop more practical versions of the ceramic and other compounds that are already available.

This means, to take one example, that for successful high-temperature superconducted power transmission, these new materials would have to have the flexibility, strength, and ductility expected of conventional metal cables. In addition, they would have to be capable of carrying adequate current densities.

It may well be that the most important applications of high-temperature superconductivity are quite unforeseeable at this moment in time. Heike Kamerlingh Onnes, when he discovered low-temperature superconductivity in 1911, could hardly have imagined that one of its most useful applications would be in a device that didn't even exist in his day, i.e., the magnetic resonance imager.

Japan and the United States are the leaders in the race toward practical utilization of high-temperature superconductivity, but there is a concern that the Japanese will surpass the United States in superconductivity development. The Japanese government increases its annual funding toward superconductivity research, while the United States does not.

As with many new technologies, one of the major concerns has been the slow progress of the commercial applications promised by researchers after the first findings in high-temperature superconductivity. Even though researchers claim that superconductivity will eventually touch every aspects of our life, few, if any, superconductivity applications have appeared, which frustrates both researchers and the public alike. Re-cent cutbacks in government funding haven't helped either. One of the biggest setbacks was when a planned superconducting facility called the Superconducting Supercollider project, which was scheduled for construction near Waxahachie, Texas sometime in the 1990s, was canceled.

All other advanced countries are also engaged in this effort; the next few years should see some further developments. Despite the slower-than-expected progress, researchers still believe that advances in superconductivity will be associated with energy savings, cleaner environment, faster transportation, and speedier computers in the future.

FURTHER READING

Books

Breakthrough: The Race for the Superconductor. Herbert M. Hazen. New York: Summit Books, 1988.

An interesting, behind-the-scenes narrative of the efforts of a small team of scientists to find a practical high-temperature superconductor.

High-Temperature Superconductivity: Experiment and Theory. Nikolai M. Plakida. New York: Springer-Verlag, 1995.

An extremely technical overview of high-temperature superconductors.

Superconducting Levitation: Application to Bearings and Magnetic Transportation. Francis Moon. New York: John Wiley & Sons, 1994.

Introduction to the basic principles of levitation using superconducting materials.

Superconductivity: The Next Revolution? Gianfranco Vidali. Cambridge, England: Cambridge University Press, 1993.

A general overview of the history and future of superconductivity.

Articles

"Antigravity? Well, It's All Up in the Air." O. Port. *Business Week,* February 17, 1997, 97.

Superconducting and antigravity (Meissner effect) progress.

"Probing High-Temperature Superconductivity." J.R. Kirtley, C.C. Tsuei. *Scientific American,* August 1996, 68-73.

An interesting look into the latest in high-temperature superconductivity.

SECTION TWO
High-Tech Principles and Concepts

Modern technology consists of a vast array of hardware and software systems built around a relatively small number of basic principles and concepts. This section presents an overview of the most important of these principles and concepts.

The single most critical factor in comprehending the new technology is to gain some understanding of the electromagnetic spectrum. All high-tech devices depend, to a greater or lesser extent, on electromagnetic energy. This section begins with a discussion of the spectrum and then moves on to high-tech communications, a primary user of electromagnetic energy.

Digital signals, in the form of binary numbers, are rapidly becoming the language of high-tech communications. The metric (SI) system is the measurement tool of the high-tech world. All of these concepts are covered in this section.

The Electromagnetic Spectrum

BASIC DESCRIPTION

The electromagnetic spectrum is made up of a group of energetic waves that range in a continuous gradation from extremely long radio waves to very short, high-frequency gamma rays. Within its range, the spectrum includes radar, TV, microwaves, infrared, visible light, ultraviolet, and X-rays.

GENERAL COMMENTS

We are all familiar with light, and many of us know that it travels at a speed of about 186,000 miles per second. Some of us are also aware that, according to Einstein's theory of relativity, no moving body can ever exceed that speed. It's not as well known, however, that light is a close relative of radio, and that radio, TV, microwaves (the same ones used to heat snacks), visible light, X-rays, and gamma rays are all members of the same family and share two fundamental family traits:

1. They can travel through a vacuum.
2. They travel at the rate of approximately 186,000 miles per second, in a vacuum.

Some scientists have suggested that all members of this family could be referred to as "light" because they so closely resemble each other in important ways. However, the family has been officially designated the "Electromagnetic Spectrum."

The spectrum is made up of energetic waves (for the moment we can simply think of waves as ripples on the surface of a pond) that are emitted when a charged particle, such as an electron (+), is excited or accelerated by a force. We tend to think of waves as motion caused by a force in some kind of medium, such as the motion of water in the form of waves caused by the wind. However, electromagnetic waves do not require any medium (atmosphere or water), because they can pass through a vacuum.

In fact, electromagnetic waves seem to behave at times as if they were particles, i.e., tiny physical bodies.

This puzzling fact has bothered physicists ever since the spectrum was discovered; the current approach is to simply accept them as either waves or particles, depending on the context in which they are being studied. The electromagnetic particle, incidentally, is referred to as a "photon" and is a tiny little weightless packet (or quantum) of energy.

The Earth depends for its existence on electromagnetic energy, mostly in the form of heat and light coming from the Sun, and a cutoff of that energy would mean the immediate end of our world.

BACKGROUND

Michael Faraday (1791-1867), whose brilliant discoveries in electromagnetism (he discovered the principle of the electric motor and invented the electric generator [+]) are landmarks in the development of modern physics, was the son of an English blacksmith and received only a few years of schooling. Because of his lack of formal training (particularly in mathematics), Faraday was not able to provide a theoretical foundation for his discoveries. His inquiring mind and his unequaled experimental skills brought him many of the right answers about how electricity behaves; but he could not always explain them. (It's interesting to note that this great unschooled genius had been "turned on" to science by reading a series of books called *Conversations on Chemistry*, written by a contemporary science popularizer named Jane Marcet. That Marcet's skillful presentation of her material was a significant factor in bringing Faraday into the scientific fold can hardly be doubted.)

Some time during the 1860s, James Clerk Maxwell (1831-1879), a Scottish physicist and mathematician, resolved to provide a theoretical foundation for Faraday's discoveries. As Maxwell developed equations to describe the behavior of electromagnetic phenomena, he discovered to his astonishment that his equations were telling him that all known electromagnetic radiation travels at the same speed as light, i.e., about 186,000 miles per second. His logical mind led him to the only possible conclusion: light itself must be a form of electromagnetism. In addition, his equations led him to predict the existence of as yet undiscovered forms of electromagnetic radiation.

As we now know, Maxwell was correct in his conclusions about the electromagnetic spectrum, and his contributions to the understanding of the physical world rank with those of Newton and Einstein. It is also extremely interesting to note that it was through the power of pure mathematical logic that Maxwell was able to disclose a previously undreamed-of relationship in the physical world.

ELECTROMAGNETIC WAVES

All of us are aware, in a general way, that powerful radio transmitters send electromagnetic waves through the air (and sometimes through space) and that these waves reach our radio sets and are converted into sound in the form of words or music.

We are also generally aware that TV signals are sent pretty much the same way, except that now much TV transmission comes into our homes through cables rather than through the atmosphere.

The Structure of Waves

We have been talking about light and other electromagnetic waves without offering any detailed explanation of these terms. First, let's consider waves. The simplest and most familiar waves are those we see in a lake or ocean. We can visualize those waves as they are sometimes drawn by cartoonists, that is, as very simple and regular:

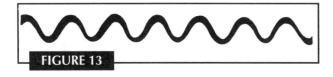

FIGURE 13

Using the concept of these simple waves, we can identify certain characteristics of all waves. The wavelength is the distance from one crest to the next.

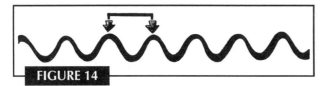

FIGURE 14

The frequency is the number of times per second the wave goes completely through its full cycle, i.e., how many times it rises to a crest, falls, then rises to a

crest again. It's likely that a huge ocean swell will have a low frequency (that is, it takes a long time to hit its crest, fall, and then rise to a crest again), whereas a small wave will have a high frequency (like a ripple).

Electromagnetic waves are much more complex than ocean waves, but their wavelengths and frequencies are determined in the same way.

Radio, TV, and radar are all close cousins (because of their frequencies and wavelengths) and are grouped near each other at one end of the electromagnetic spectrum.

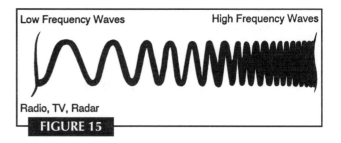

Low Frequency Waves   High Frequency Waves

Radio, TV, Radar

FIGURE 15

Visible light is found more toward the middle of the spectrum.

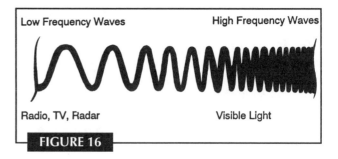

Low Frequency Waves   High Frequency Waves

Radio, TV, Radar   Visible Light

FIGURE 16

Visible light is such a fundamental factor in our daily lives that it's difficult to visualize it as a tangible entity that travels at a finite speed, and that obeys other physical laws. (While 186,000 miles per second is incredibly fast by terrestrial standards, astronomers take it in stride. It takes light from the Sun, at an average of 93 million miles away, about eight minutes to reach the Earth.)

The fact that we can see the Sun, the Moon, and other celestial bodies depends entirely on the ability of light to pass through a vacuum. (Space is essentially a vacuum—not a perfect one, but more perfect than any yet produced on Earth.) Sound, for example, cannot pass through a vacuum; it needs a medium (e.g., water or air) to carry its waves. Without the atmosphere, the Earth, for all practical purposes, would be silent.

Note: When we talk to astronauts as they circle the Earth, only electromagnetic impulses in the form of radio waves are going from us to them, and from them to us. These impulses are converted into sound by electromechanical devices (radios and speakers) up there and down here, where there is either a natural or artificial atmosphere to carry the vibrations coming from the speakers to human ears.

We can see how vital the special characteristics of electromagnetism are (its great speed and ability to go through vacuums) to the entire area of space activity and exploration, by reflecting back to the now-famous visits of the Voyager 1 and 2 spacecrafts to the outer planets. It was necessary for us to communicate with them regularly in order to control their positions with respect to the planet, and to receive the pictures that they were sending to us in the form of numbers. (See "Digital Image Processing.")

For example, when Voyager 2 was at the planet Neptune, how could scientists control an object almost three billion miles away? Only with the speed and power of electromagnetic energy. If we divide 2.7 billion (2,700,000,000) miles by the speed of light (or electromagnetic waves of any kind), 186,000 miles per second, we get 14,516 seconds. If we divide this by 60 seconds per minute, we get 241 minutes, and if we divide again by 60 minutes per hour we get a little over 4 hours. So every transmitted instruction took a little over four hours to get to the distant spacecraft. Such lengthy transmission time made things difficult, but resourceful scientists and engineers were able to make the necessary adjustments to control the spacecraft and its instruments.

Wavelengths and Frequencies

Radio waves are relatively long, with wavelengths ranging from a few miles to a fraction of an inch. Their frequencies vary from about 10,000 cycles per second up to 300 billion cycles per second. Incidentally, it's customary now to say "hertz" (Hz) instead of cycles per second; among engineers and scientists, one million cycles per second, for example, would be referred to as one megahertz or MHz, "mega" being the metric prefix for a million.

Visible light has a much shorter wavelength than radio; it's in the range of several trillionths of an inch. Its frequency (the number of times it goes through its cycle each second) is about 1,000,000,000,000,000 Hz or 10^{15} Hz.

Note: Because frequency and wavelength numbers get so high and so low, it's burdensome to write them out each time. Most scientists and engineers use a form of scientific "shorthand" to avoid the long strings of zeroes. For example, the frequency of visible light is usually given as the range between 10^{14} (read as 10 to

the 14th) and 10^{16} hertz (or cycles per second). The expression 10^{14} means 10 multiplied by itself 14 times. The little number at the right side of the 10 is called an "exponent." We can understand this usage by looking at some more familiar examples: 10^2 (10 times 10, or 10 squared) equals 100, 10^3 equals 1,000, 10^6 equals one million, 10^9 equals one billion, and 10^{12} equals one trillion. So we can see that 10^{14} is a very large number, larger than a trillion. To show three trillion we would write 3×10^{12}.

Where frequencies lead to big numbers, wavelengths (with the exception of the radio end of the spectrum) tend to be small numbers. They are expressed with a negative exponent, e.g., 10^{-14} (read as 10 to the minus fourteenth). Again we can grasp this more clearly by considering familiar examples: 10^{-2} equals one one-hundredth, 10^{-3} equals one one-thousandth, 10^{-6} equals one millionth, 10^{-9} equals one billionth, and 10^{-12} equals one trillionth. So 10^{-14} is a very small number, smaller than a trillionth.

There is no scientific necessity for using the exponential system, but as a practical matter it saves space and makes these very large and very small numbers easier to talk about. Most tables and charts of the electromagnetic spectrum will show wavelengths and frequencies in this notation. They will also generally use the centimeter for wavelengths; the centimeter is a metric (+) measurement unit and is less than half an inch in length.

Returning to the mental picture of ocean waves that we considered earlier, it isn't particularly difficult to visualize a whole range of such waves, starting with tiny, fast-cycling ripples, gradually increasing to medium-length waves, and continuing all the way up to long, majestic, slow-cycling breakers. This, in principle, gives us a mental picture of the electromagnetic spectrum. Some radio waves cluster near one end of the spectrum (they're our long majestic breakers), and fast-cycling, incredibly short gamma rays (they're our tiny ripples) cluster near the other end. In between, we find TV, radar, infrared, visible light, ultraviolet, and X-rays. (See figure 17.)

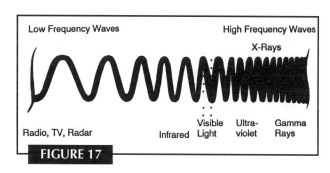

FIGURE 17

The electromagnetic spectrum is continuous; the terms "radio," "infrared," and "X-ray" are convenient labels used to identify regions of the spectrum that interact with matter in certain recognizable, characteristic ways.

X-rays, for example, can pass through some materials that block visible light completely. On the other hand, visible light can pass through the Earth's atmosphere, which largely blocks X-rays. All regions of the spectrum, however, possess the two familiar and vitally important family traits: they can pass through a vacuum, and they move at 186,000 miles per second, in a vacuum.

All modern technology, all communications, all production and transportation, all bodily functions of living creatures, and, in fact, the very existence of the planet are dependent on electromagnetic energy. We can do without gravity, one of the other fundamental forces that we are familiar with, for extended periods of time, as proved by astronauts in the weightless environment of space. However, we could not survive without electromagnetic energy for more than a split second.

Where does this energy come from and how are we able to harness it so effectively? To answer this question we need to review our understanding of atoms and electrons, because events at the atomic level are the source of this energy.

Atoms and Electrons

Most people are aware (at least in a general way) that matter is made up of atoms, which consist essentially of a central nucleus and one or more orbiting electrons. Atoms are about one-hundred millionth of an inch in diameter.

An electron is a tiny, stable particle with an extremely small but measurable mass. It bears a negative electric charge. (A charge is a force that can be exerted on another charged substance. Two substances with the same charge will exert a repelling force on each other; substances with opposite charges will exert an attractive force on each other.) The terms "negative" and "positive" are only arbitrary conventions (they could be switched around without violating any principles), but as currently used, they provide an accepted means of classifying charged substances.

The nucleus of an atom is made up of one or more protons and one or more neutrons. (The hydrogen atom excepted: it has only a proton.)

A proton is (like the electron) a tiny, stable particle but it bears a positive charge and has a much greater mass than does the electron. The positive charge of the proton has exactly the same force as the negative

charge of the electron, so that when they act on each other they neutralize each other's charge. Neutrons have a slightly greater mass than protons and are electrically neutral, i.e., they have no charge.

The commonly shown model of the atom looks like a little solar system and is a good enough representation for our purposes. (See figure 18.) Particle physicists have developed a much more complex picture than this: one in which the electrons are viewed more as clouds diffused into specific energy bands; and the nucleus itself is envisioned as a complex structure in which protons and neutrons are made up of even smaller building blocks of matter.

The number of protons in the nucleus determines the identity and chemical properties of a particular atom. Hydrogen has one proton, helium has two, lithium has three, and so on. (Some scientists have suggested that it would have been much simpler to describe the elements by their number of protons, .i.e., Element I for hydrogen, Element II for helium, and Element III for lithium, and so on, rather than using the elaborate names and abbreviations now assigned to them.)

While neutrons do not substantially affect the chemical properties and reactions of an atom, they become extremely important in physical and nuclear reactions.

AN IDEALIZED SCHEMATIC REPRESENTATION OF THE OXYGEN ATOM. IT HAS EIGHT PROTONS AND EIGHT NEUTRONS PACKED INTO ITS NUCLEUS, AND EIGHT ELECTRONS SURROUNDING THE NUCLEUS (TWO IN AN INNER ORBIT AND SIX IN AN OUTER ORBIT).

FIGURE 18

The relationship between an atomic nucleus and its electrons is not always a faithful one. Electrons are lost, regained, exchanged, and shared with other atoms; much of the subject matter of chemistry, in fact, deals with the compounds and reactions that come into being when atoms share or capture the electrons of other atoms.

When an atom loses or gains electrons (which may occur because of outside forces or changes in temperature), its charge is no longer neutralized, and it is referred to as an "ion," i.e., a charged atom. The escaped electrons, if they are not recaptured by another atom, also now exert an active force in the form of a negative charge.

Bodies of living creatures are composed of atoms that conform to the same physical laws that govern the atoms of nonliving substances. When high-energy electromagnetic waves such as X-rays and gamma rays penetrate living tissue in excessive doses, they ionize (rip away electrons) from cellular atoms and upset the electrical balance of the organism. They may also penetrate the chromosomes (the gene-bearing bodies of living creatures) and damage the genetic structure of the organism. For this reason, the high-energy, upper-frequency range of the spectrum is referred to as "ionizing radiation." (Recently, a good deal of concern has been expressed about the possible long-term effects of low-frequency electromagnetic radiation, which had long been considered safe. Ongoing studies are now evaluating this potential hazard, but no damaging evidence has yet been found.)

Electrons and Electromagnetic Energy

When electrons (or any charged particles) are accelerated by some energy input such as heat, light, or a voltage, they emit photons. Photons are massless particles that form the wave structure of electromagnetic radiation. Photons are the smallest unit (or quantum) of electromagnetic energy.

If the photon has a low frequency when it is emitted, it will form a wave in the low-energy part of the spectrum, such as radio or microwaves; if it has a high frequency, it will form visible light or X-rays or gamma rays.

The frequency of the photons is determined by the energy level of the charged particles that emit them; in many human-engineered situations such as radio broadcasting, the desired frequencies can be obtained by exciting electrons in oscillators, which vibrate consistently at predetermined energy levels (or frequencies).

When we turn the light switch on in a dark room, the heat generated by the electric current excites the electrons in the lamp's filament, causing them to emit photons. A 60 watt bulb will emit about 10^{20} photons per second. Some of the photons will oscillate in the 10^{14} to the 10^{16} hertz range, thereby providing visible light for human eyes.

Note: It's interesting to reflect that if our switch were in San Francisco and our light bulb in New York, it would light up almost as quickly when we pulled the switch as if everything were in the same room. It isn't electrons traveling through the wire rapidly that bring this about. They are moving rapidly but much below the speed of light. In an AC (alternating current) circuit, in fact, they are moving back and forth, always trying to head toward the positive terminal of the constantly changing circuit. (See "Electrical Power Generation.") It's the electromagnetic wave, formed in the conductor by the emission of photons from the voltage-excited electrons, that brings the near instantaneous response to the activation of the switch. The electromagnetic wave does travel through the conductor at near the speed of light.

Every minute of the day or night, no matter where we are or what we are doing, we are surrounded by electromagnetic radiation, which comes from either human-engineered devices or from natural sources. However, unless we have some kind of correctly tuned receiver such as a radio or TV set, we are not consciously aware of this radiation, with the exception of visible light (and infrared rays, which we may perceive as heat). Our eyes are correctly tuned for the 10^{14} to the 10^{16} hertz frequency of light. The pattern carried by the light waves will be transmitted through our eyes, across our internal nerve network to our brain, where an image will form in a much shorter time than it takes to read these words.

The rather narrow, undistinguished region of the electromagnetic spectrum between frequencies 10^{14} and 10^{16} would be little more than a crossover junction between infrared and ultraviolet, if it weren't for the living creatures of the world, who perceive it as light.

Summary

- The electromagnetic spectrum is made up of a group of energetic waves that range in a continuous gradation from long, low-frequency radio waves to short, high-frequency gamma rays. Within its range, the spectrum includes radar, TV, microwaves, infrared, visible light, ultraviolet, X-rays, and gamma rays.

- All electromagnetic waves can pass through a vacuum, and all travel at the approximate rate of 186,000 miles per second, in a vacuum.

  The spectrum is continuous, and different waves gradually change their characteristics as wavelength and frequency change. The names X-ray, ultraviolet, and so on describe loosely bounded regions of the spectrum.

- Different regions of the spectrum interact differently with matter. For example, X-rays can penetrate some materials that block visible light; however, visible light can pass through the Earth's atmosphere, which blocks X-rays.

- In the high-frequency regions of the spectrum, starting in the ultraviolet region and continuing through X-rays and gamma rays, electromagnetic waves become so energetic that they can seriously damage the biological systems of humans and animals. This region of the spectrum is referred to as ionizing radiation.

- Electromagnetic energy is essential to the existence of the world as we know it. Matter could not retain its present form, and life could not survive for even a fraction of a second, without it.

FURTHER READING

Books

Electromagnetic Concepts and Applications. Richard E. Dubroff, S.V. Marshall, G.G. Skitek. New York: Prentice Hall, 1996.

In-depth and applications-oriented approach to introductory electromagnetism.

Electromagnetic Fields: A Consumer's Guide to the Issues and How to Protect Ourselves. B. Blake Levitt. New York: Harvest Books, 1995.

A resource mainly for consumers, but a good explanation of electromagnetic fields.

Electromagnetism. John C. Slater, Nathaniel H. Frank. New York: Dover Publications, 1969.

Good basic coverage of the subject.

Maxwell on the Electromagnetic Field: A Guided Study. Thomas K. Simpson, Anne Farrell. New Jersey: Rutgers University Press, 1997.

A compendium of James Maxwell's notes and their interpretation.

Articles

"Detecting Natural Electromagnetic Waves." S. Carlson. *Scientific American*, May 1996, 98.

A synopsis of natural electromagnetic waves.

"Electromagnetic Fields: No Evidence of Threat." National Research Council. *Consumer's Research Magazine*, December 1996, 25-27.

A discussion of the health concerns and findings about low-frequency electromagnetic radiation.

Communications

BASIC DESCRIPTION

Communications is the process of transmitting and receiving information in the form of signals; included among the vast array of commonly used signals are words, gestures, and electromagnetic patterns. High-tech communication uses electromagnetic waves and pulses in varying patterns to send and receive information. Some of the most dramatic changes in modern society have resulted from the ability to send near instantaneous messages and images from one place on earth to another, thousands of miles away. International soccer matches and political upheavals have become global events, watched by hundreds of millions of people at the same time, through the controlled use of electromagnetic signals.

Note: Four common words—"data," "information," "signal," and "message"—are used throughout this section. The following informal definitions should help to clarify their use:

- data—facts, observations, or statistics
- information—meaningful data
- signal—a unit of information
- message—one or more related signals

BACKGROUND

When two people meet and engage in conversation, they are exchanging signals in the form of spoken words (or in the case of deaf people, in the form of signs). When we stop for a traffic light, we are responding to a visual signal in the form of a color. When a beaver's tail slaps the surface of a pond, a danger signal is being sent to the rest of the family, in the form of a loud noise.

Living creatures develop ways of communicating with each other in seemingly endless patterns of signals.

In the high-tech world, however, systematic, high-volume, long- and short-distance communication is accomplished by utilizing electromagnetic signals in one of the following ways:

- Electromagnetic waves sent through the atmosphere or through empty space—through-the-air systems
- Electromagnetic waves sent through a waveguide or conductor such as fiber optics(+) cable or metal cable

In practice, many modern communications systems are hybrids and use both methods in transmitting a single message.

In general, the radio end of the spectrum is favored for "through-the-air" transmissions when the goal is to "broadcast" a signal, i.e., when a signal is intended for multiple receivers. This follows from the fact that some radio frequencies can bend around or pass through physical obstacles, such as walls. While higher frequencies, such as those of light, are also used for through-the-air transmission, they are more likely to be directed at a single receiver or a group of receivers within their "line of sight."

Radio signals are also sent through conductors and, increasingly, visible light signals are conducted through fiber optics cables.

Through-the-Air (or Through-Space) Systems

Examples: radio, TV, radar, garage door openers, remote TV controllers, pagers, traffic lights, cordless telephones, cellular phones, and lighted signs.

Advantages of Through-the-Air Systems

1. The signal transmitter and the receiver do not need to be physically connected to each other. This makes communication possible with remote sites and with rapidly moving vehicles.

2. Signals can be sent (broadcast) to a large number of receivers at the same time, as long

as they are within range of the transmitter. This makes the rapid dissemination of emergency messages possible.

3. It's possible to move transmitters (and receivers) without laying new connecting lines.

4. It's possible to reach an individual or vehicle whose whereabouts is not known. Pagers and cellular phones enable some organizations (for better or worse!) never to be "out of touch" with their key people.

5. It is very helpful in the study and protection of wildlife. By placing a small radio transmitter on an animal, experts can learn a good deal about the animal's life cycle without being intrusive.

Limitations of Through-the-Air Systems

1. Because there are a limited number of broadcast frequencies available, the use of transmitters must be restricted in some way or they will seriously interfere with each other.

2. Through-the-air signals can be distorted by atmospheric disturbances, and through-space signals are sometimes bothered by an increase in solar activity.

3. Physical obstacles such as mountains, bridges, or buildings can prevent or interfere with some transmission.

4. Through-the-air signals are not private: they can be picked up by any correctly tuned receiver within their range. This lack of privacy is a significant problem with cordless and cellular telephones. (It's true that cellular messages can be "scrambled" by encoding devices at the transmission point and "unscrambled" at the receiving end, but this feature is an added cost and an added complication, and it will not necessarily defeat a really determined "eavesdropper" who may use counter-effective electronics to decode the message.)

Conducted Signals

Examples: Fiber optics or copper-wired telephones, cable TV, wire-connected computer elements, wired intercoms and microphones, undersea fiber optics, or copper cable systems.

Advantages of Conducted Signals

1. Signals can be directed very precisely through conductors.

2. Conducted signals are much more private than through-the-air signals; one must "tap" into the line to intercept a message.

3. They are far less vulnerable to atmospheric disturbances.

4. They have greater message-carrying capacity because the same frequencies can be used in close proximity to each other.

5. They are not blocked by physical obstructions; they can go anywhere their conductors can reach, e.g., inside the human body or through massive structures.

Limitations of Conducted Signals

1. The fundamental limitation of conducted signal systems is the need for a physical connection between the transmitter and the receiver. Conducted messages cannot be sent to, or received from, moving cars, airplanes, or space vehicles.

2. Laying the connecting lines between message centers can be a labor-intensive, high-cost process. Systems builders get involved with right-of-way clearances that can sometimes lead to litigation. And, invariably, there will be time-consuming encounters with administrative bureaucracies.

3. The maintenance of installed lines can also be a labor-intensive, high-cost process and is often carried out under pressure when users are deprived of service. The existence of long cables between transmission and receiving points increases the possibility of accidental damage, as well as deliberate sabotage or vandalism.

Metal Cable Conductors Compared with Fiber Optics (+) Cable

The traditional signal conductor has been metal cable; the relatively recent appearance of fiber optics has dramatically affected the communications field.

Metal cable systems generate a signal-carrying electromagnetic wave by imposing a force in the form of a voltage on electrons in the conductor. Fiber optics systems send laser- or diode-generated (+) light signals through tiny strands of glass or plastic.

Because the light waves have a higher frequency than the voltage-generated waves, their message-carrying capacity is much greater. (They can carry more "fine print," so to speak.)

Although there are many other advantages to fiber optics the bottom line is its message-carrying capacity.

In telephone applications, for example, approximately 8,000 simultaneous conversations can take place on a fiber optics wire as compared to 48 on a copper wire.

Today's Wireless Society

Four concepts have created excitement in the communications world: digital signals (+), fiber optics (+), networking (+), and wireless telephones. Since the first three topics are covered elsewhere in this book, we will confine ourselves here to discussing the fourth big factor in today's (and the future's) communications, the wireless telephone.

Wireless telephones, which include cellular phones and cordless home phones, are really integrated radio transmitter-receivers. What makes them different from the familiar "walkie-talkies" or CB radios that have been broadcasting signals to each other for decades? The genius of the cellular telephone is that it allows an ordinary radio (not much different from a walkie-talkie) to hook up to the telephone network. Unlike walkie-talkie users, who can only communicate with each other, and only within a limited range, the cellular user can harness the power and "reach" of the vast web of existing telephone connections and talk to just about anybody in the world who has a telephone.

The radio transmitter-receivers, which we know as cordless, cellular, briefcase (transportable), and pocket telephones, are built in the familiar shape of conventional telephones and have the kinds of controls that we're used to. But, as indicated, it is only their ability to gain access, through their frequencies and codes, to stations that are hard wired to the telephone network that differentiates them significantly from conventional radio transmitters.

In the case of cordless home phones, the receiving station is in a box hanging on the kitchen wall or located somewhere else in the house. In the case of cellular phones, the station is a "cell," each one a small hexagonal area with a receiving-transmitting base station strategically located to serve the needs of mobile clients.

The only limitations on the mobile client are the distance from a "cell," the amount of "traffic" (or other clients on their cellphones) on the cell, and the amount of power available to the mobile transmitter. The client must depend on a battery for power, and while a fairly large battery or an AC adapter can be used in a car, a briefcase or pocket telephone user must get by with a smaller battery.

Summary

In its simplest form, the communication system has a signal sender (called a transmitter), a path (air, space, or a conductor) and a signal-detector (a receiver).

The signal process usually begins with some mechanical action, such as a person speaking into a telephone or striking the keys of a keyboard. (See Figure 19.) This mechanical signal is converted into an electrical impulse and carried on an electromagnetic wave, over the path, to the receiver. There, it is (usually) converted back into mechanical form such as a loudspeaker's reproduction of the original speaker's voice, or a printed copy of the message from the keyboard. (In the case of a garage door opener, the signal activates a motor that opens or closes the garage door.) The signal may also appear as a visual message on a CRT (+) screen, or in many other forms. Two-way systems have transmitters and receivers at both ends, and passage through the path may become more complicated, but the fundamental structure remains the same.

No matter how complex the message being sent, no matter whether the signal path is a tiny fraction of an inch or millions of miles, and regardless of the intricacy of the signal-sending device and the receiver, the basic pattern shown in figure 19 will be followed.

THE FUTURE

Unquestionably, wireless telephones have vastly enlarged the "reach" of existing communications systems. Digital signals and fiber optics will both contribute to an enormous increase in the amount of information that can be handled on communications systems. As more optical fiber (+) cables are laid down to enhance communication lines, communications will become faster, and new technologies will develop as a result, especially in telephones and computers, such as capacity for higher modem speeds. In addition, several new satellite communication networks have been proposed, including a worldwide system based on transmitters in space. The system will include about 77 satellites, all uniformly spaced about 475 miles above the surface of the Earth. This vast array of satellites will link digital signals to form a cellular network, allowing clients on the system to communicate with any other person on the phone network.

Wireless communications are not the only players in the changing world of communications. With the popularity of such computer communications as the Internet, networking will (and already has) fundamentally change the way people communicate. (See

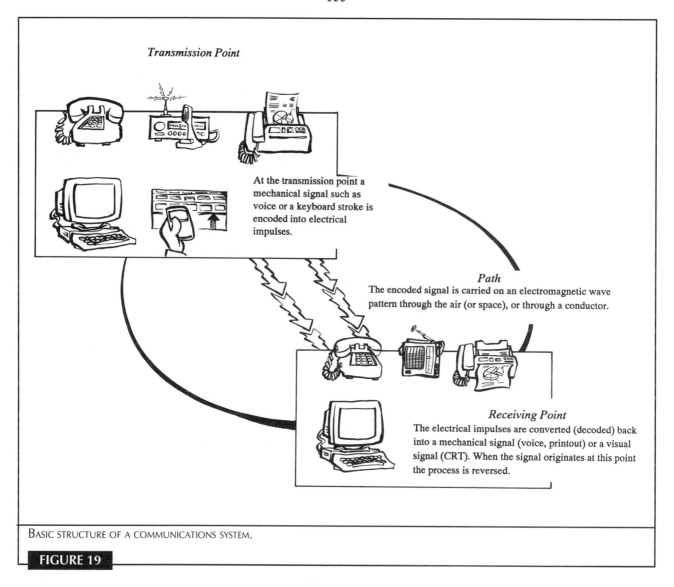

Transmission Point

At the transmission point a mechanical signal such as voice or a keyboard stroke is encoded into electrical impulses.

Path
The encoded signal is carried on an electromagnetic wave pattern through the air (or space), or through a conductor.

Receiving Point
The electrical impulses are converted (decoded) back into a mechanical signal (voice, printout) or a visual signal (CRT). When the signal originates at this point the process is reversed.

BASIC STRUCTURE OF A COMMUNICATIONS SYSTEM.

FIGURE 19

"Networking.") Although fiber optics transmission lines allow a greater number of calls over the phone lines, the biggest problem in such communications is on the horizon: Trying to increase the bandwidth of the communication systems to allow the projected larger amounts of traffic to successfully communicate over the lines. This problem has already been demonstrated during major events: Personal computers attached to communication systems on the Internet often are "shut out" of the system as hundreds of thousands of people also try to enter the system.

Far into the future, some researchers believe we may be able to use photons, electrons, or other quantum particles to carry information. Quantum communications would be faster as each particle would best carry a piece of information; so far, no one knows *how* the particles would carry the data. Plus, scientists are still puzzled about how to code and decode the information and how to use as little energy as possible to carry the information.

FURTHER READING

Books

Cable Communications: Building the Information Infrastructure. Eugene R. Bartlett. New York: McGraw Hill Text, 1995.
Describes the technology that makes up the information superhighway, with an emphasis on cable systems.

Communications Media in the Information Society. Joseph D. Strubhaar, Robert Larose. New York: Wadsworth Pub. Co., 1996.

A look at the communications media of today and the future.

Future Talk: The Changing Wireless Game. Ron Schneiderman. New York: IEEE, 1997.

A good nontechnical synopsis of global wireless communications and its impact on our future.

Telecommunications Primer: Signals, Building Blocks, and Networks. E. Bryan Carne. Englewood, NJ: Prentice Hall, 1995.

Comprehensive overview of modern telecommunications.

Articles

"Cellular Goes Digital." Dawn Stover. *Popular Science*, January 1990, 52-55.

The arrival of digital cellular phones.

"The Elbowing Is Becoming Fierce for Space on the Radio Spectrum." Keith Bradsher. *New York Times*, June 24, 1990, 1.

Discusses the competition for ever-scarcer radio frequencies.

"How the World Became Smaller." Mary Brooke. *History Today*, November 1996, 45-50.

A brief history of communications.

Digital Signals

BASIC DESCRIPTION

A digital signal is a unit of information expressed as a number or as a numerical representation. Digital signals use a limited number of sharply defined symbols to convey messages and to describe physical objects and values. Because of this arrangement, information transmitted in digital form is virtually free of interference and background noise and is easily copied and reproduced. Increasingly, high-tech communications systems are sending their messages in digital form.

BACKGROUND

Two major techniques for processing and transmitting information are in use: analog and digital systems. The two systems are described below:

1. Analog Signals: Analog signals use electromagnetic waves, physical forces, objects, or distances to proportionally represent the measurements and attributes of an entity and to transmit messages. Analog signals may take an unlimited number of shapes or forms.

2. Digital Signals: Digital signals represent information in numbers. Digital systems use only a limited set of discrete symbols to describe all of the attributes and measurements of an entity and to transmit messages. In the high-tech world, binary numbers (+) are the most commonly used digital signals.

Following are some common items listed in their analog and digital counterparts.

Analog	Digital
watch with hands	watch that shows numbers
slide rule	calculator
LP record	compact disc
photograph	digitized image (+)

Keeping Time

An analog watch represents the passage of time by causing tiny metal (or plastic) hands to travel around the face of the watch; i.e., the passage of time is measured by the distance that the hands (usually representing the passage of hours, minutes, and seconds) have moved during an interval. At any particular moment, one reads the time by noting the position of the hands on the watch.

In a digital watch there are no hands; the time is shown as an arrangement of numbers. The passage of time is represented by the orderly changing of the numbers.

The analog watch measures time continuously. The digital watch represents time in discrete (clearly separated) jumps, from one number to another.

The mechanism for producing the visible time signal, which either watch provides, may vary radically or may be very similar. Some watches are still operating on a spring-driven mechanical configuration, while others (perhaps most) are now using precisely vibrating crystal devices and are driven by batteries.

It should be noted that it is quite possible for an analog watch to be more accurate than a digital watch. The fact that a system provides a digital readout has little to do with its accuracy.

However, if a particular digital watch is at least as accurate as a particular analog watch, the former possesses a significant advantage over the latter, as shown in the following simple imaginary experiment: We take two watches of equal accuracy, one analog and one digital, and stop them at the same instant. Let us say that they read 3:32 at that instant. We then ask someone to write down the time from each watch. In the case of the digital watch, the numbers are to be copied. With the analog watch, the person is to draw the watch face with the hands correctly positioned. The two copies are then handed to another person who will copy the copies. If this process is repeated, let us say, 10 times, it will soon become obvious that the most recent analog copy, i.e., the copy showing the positioned hands on the watch face, will have diverged significantly from what was on the original watch. On the other hand, the most recent copy made from the digital watch will, in all probability, still show 3:32.

The point is that it's much easier to copy numbers accurately than it is to copy patterns. This is precisely what happens in the real world of signal processing. It's more difficult for sophisticated communications equipment to faithfully transmit complex wave patterns than to transmit numbers in the form of discrete pulses. In almost every processing step, analog signals tend to degrade and become distorted to some extent, while digital signals retain their integrity.

Recording Music

The information stored in the grooves of an LP record is an analog of the recorded music. During the recording process, the variations in tone, volume, and other characteristics of the music have caused deflections in the path of a stylus, which has been tracing grooves in a soft master recording medium.

These patterns of deflections in the LP grooves are proportional to the sound waves emanating from the musical instruments. Later, when hardened copies of the master are available, playback machines will reverse the recording process, using a stylus that is guided by the deflection patterns in the hardened grooves of the completed disc. The physical movements of the guided stylus are monitored and converted, first into electrical impulses and then into sound, providing a realistic representation of the original music.

A digital recording, on the other hand, represents its musical sound in numbers. An acoustical sampling device assigns numbers to sound waves on the basis of their musical characteristics. In concept, this procedure is very similar to the way an optical scanner assigns numbers to a photograph. (See "Digital Image Processing.") Now the finished disc, instead of having grooves etched by stylus deflections as in the LP, will have only numbers on its surface. In practice, the numbers will be in the form of tiny dots or blanks, representing binary digits (+).

When the disc is played, a laser (+) "reads" it by sweeping a tiny spot of light over the surface and monitoring the reflections. The information read this way will be converted into electrical impulses, and then into musical sound.

An important difference between the two systems is that, during the playback process, the compact disc (digital) system will accept, as signals, only precisely delineated, exactly positioned dots or blanks on the surface of the disc, while the LP (analog) system may accept accidental groove deflections as legitimate signals. The CD system is not easily fooled by a speck of dust. It will require gross defects in the disc to cause distortion in the sound produced. The LP system, on the other hand, is easily deceived. Because its driving signal is so variable, it has trouble discriminating between the wanted signal and unwanted noise. It tends to accept the stylus deflections caused by specks of dust, or chipped grooves, as genuine signals, reproducing them as pops and crackles through the loudspeaker.

It should be noted, however, that it is quite possible for a digital (CD) recording to be inferior to an analog (LP) recording because of poor sound sampling or processing, or poor output equipment. Both digital and analog long-distance telephone lines, for example, have speakers at both ends. A well-transmitted analog or digital signal could be ruined by a poor voice reproduction mechanism. Thus, both digital and analog systems are at the mercy of their input-output devices and other characteristics of the network.

In all of the excitement over digital systems, it should not be forgotten that analog systems have provided, and will continue to provide, very satisfactory performance. LP recording devices have contributed outstanding musical performance for many years. The analog watch is still preferred by many people, because they have become accustomed to visualizing time in terms of the spatial relationships on the face of the timepiece, rather than as numbers. In many engineering procedures, analog signal systems remain essential. In numerous applications it is necessary to convert from digital to analog signals, or from analog to digital, on a routine basis. And many processed signals intended for human beings (such as voiced telephone messages) must be delivered in analog form.

Summary

Both digital and analog signals are surrounded by noise. The complex pattern of analog waves makes it difficult to reproduce them accurately and to distinguish them from noise, particularly when they have become faint. Discrete, precisely delineated digital signals, on the other hand, are easily extracted from the noisy environment, even when they have become weak.

Signals are processed intensively in the high-tech world. As they move from one state or level to another, it is often necessary to copy (or regenerate) them. Each time this replication is done, the analog signal loses a little of its integrity, while the digital signal remains intact.

THE FUTURE

A significant factor in the growing importance of digital systems is that the ever-more-prevalent digital computer has an insatiable appetite for numbers and wants nothing else. Whether it be a photograph, an encyclopedia, an engineering drawing, or a chest X-ray, the computer will store it, process it, or transmit it only if it is presented in the form of numbers.

The high-tech world, in deference to the computer and to the need for capacity and accuracy in communications systems, will increasingly fit its descriptions of objects and concepts into varied arrangements of the two symbols of the binary number (+) system.

FURTHER READING

Books

The Art of Digital Audio. John Watkinson. New York: Butterworth-Heinemann, 1994.
   An introduction to the field of digital audio.
Being Digital. Nicholas Negroponte. New York: Knopf, 1995.
   Examines the implication of the digital revolution and how it's transforming the way we live.

Articles

"Digital Signals from the Sky Enliven Radio." Hans Fantel. *New York Times,* April 15, 1990, H1.
   The growth of digital signal radio in Europe.

Binary Numbers

BASIC DESCRIPTION

Modern number systems use only a limited group of symbols to denote all numbers. This system is possible because the position of a digit (any single character), as well as its rank, defines its value within a number. The binary system uses only two symbols to denote all numbers. It is the basic number system of the digital computer. The binary system is particularly suitable for computer applications because so many of the computer's operations depend on the "state" of its internal components, i.e., whether they are "on" or "off" at any particular instant. Either of these two states can be precisely described by one of the two binary symbols. (For example, "0" for "off" and "1" for "on.") An ordinary light bulb is an example of a binary system. It is either on or off. A single bulb could make a very effective (but slow) signaling system. If we set a time scale so that the light had to be on or off at any second, we would have a complete system. If the light bulb were on for one second, that would signal a one, if it were off for a second, that would be a zero. If were on for three consecutive seconds, that would be three ones, off for two consecutive seconds, two zeros, and so forth. Within the computer, millions of such changing combinations of instantaneous on-off states, within its logic switches, form the basis of its capacity to calculate and process data.

Note: In this section we will use the following informal definitions:

number—a symbol denoting a quantity of units.

decimal number system—a system that uses 10 symbols to denote all numbers. This is the most commonly used system throughout the world.

BACKGROUND

Decimal Number System

We tend to think of numbers in cycles of 10, because we live in an environment where the decimal counting system is used almost exclusively in everyday affairs.

However, it is no more necessary to calculate and to solve arithmetic problems exclusively in the decimal system, than it is to express ourselves exclusively in the English language.

There are, in fact, many other number systems which are, or have been, in use throughout the world.

Our own system of counting is called the decimal system because it rises (or descends) in blocks of 10. It utilizes 10 familiar symbols:

0, 1, 2, 3, 4, 5, 6, 7, 8, 9

These 10 symbols, in various arrangements, enable us to write any number, no matter how large or how small. In order to count beyond nine, we simply begin to repeat one rung higher:

9 + 1, 9 + 2, 9 + 3 . . .

which, for convenience, we write as:

10, 11, 12 . . .

but we understand that the position of any digit in the number governs that digit's value. In the number 2314, the 2 really means 2000; the 3 means 300; the 1 means 10; and the number could (logically but clumsily), be written: 2000 + 300 + 10 + 4. Thus, every digit has a positional as well as a numerical value.

So we advance in cycles of 10 and obtain larger numbers by appropriate positioning of our basic 10 symbols. But it is only by convention that we use 10 as our number base. Many experts feel that this derives from the age-old practice of counting on our 10 fingers. It is quite possible to count and do arithmetic using other number bases.

Other Number Systems

A number system that repeats in cycles of five requires only five symbols:

0, 1, 2, 3, 4

If we count to 10 (decimal base) in the base five system, we will have:

Decimal	Base five
1	1
2	2
3	3
4	4
5	10
6	11
7	12
8	13
9	14
10	20

The key to both of these number systems (and to all conventional number systems) is the symbol "0" (zero). It really acts as a null position marker, telling the counting system that when it has been reached it's time to start the cycle again.

Aside from its unfamiliarity, the base five system is a perfectly acceptable way of counting and calculating, and it has, in fact, been used as the basic number system of some societies. A small disadvantage is that it requires more digits to express large numbers than the decimal system does, because it uses a smaller pool of symbols. For example, the decimal number 25 would require three digits in base five, i.e., it would be 100.

If a number system with a base larger than 10 is used, it requires additional symbols. For example, a base 12 system would require two additional symbols, which we could call "A" and "B" in the following sequence:

0, 1, 2, 3, 4, 5, 6, 7, 8, 9, A, B

The base 12 system, like the base five, provides a perfectly acceptable counting and calculating system and, like the base five, has actually been used by some societies.

The Binary System

The binary system is a base two system and consequently requires only two symbols:

0, 1

If we count from 1 to 10 (decimal), the corresponding binary numbers will be:

Decimal	Binary
1	1
2	10
3	11
4	100
5	101
6	110
7	111
8	1000
9	1001
10	1010

Since it is restricted to two symbols, the binary system (as the above scale shows) requires many more digits to express large numbers than the decimal system does. Although it's easy to perform arithmetical operations in the binary system, the long strings of zeros and ones are difficult for the human eye and mind to cope with.

The binary system does, however, have many advantages for computer use. Since computers are electronically based, it is possible to express numbers as the simple absence (0) or presence (1) of a charge, or the "on" or "off" state of a component. Also, large strings of digits don't bother the computer at all.

In computer terminology, a binary digit is called a "bit." By convention, seven bits represent a character (such as a letter of the alphabet), and an extra bit is added as a check. These eight bits are called a "byte."

Although the computer does its work in binary numbers, it converts its output to decimal notation for the convenience of its human users.

In communications systems the two symbols of the binary system can easily be represented by the presence or absence of an electric or light pulse.

When engineers and scientists refer to digital systems, they are usually referring to binary digital systems.

Summary

Binary numbers are important in the new technology world because:

- They are the basic building blocks of logical computer operations.
- Their two symbols parallel the characteristics of many natural phenomena: negative/positive, high/low, on/off, dark/light, present/absent, etc. Thus they provide a powerful descriptive notation for many physical phenomena.

FURTHER READING

Books

Fundamentals of Mathematics. William M. Setek, Jr. Encino, CA: Glencoe Pub. Co., Inc., 1979 (2nd edition).

Contains an excellent section on nondecimal number systems.

Mathematics for Computer Science. A. Arnold, Irene Guessarian. Englewood, NJ: Prentice Hall, 1996.

A good cross-section of mathematics and computers, including a chapter on logic.

The Metric (SI) System

BASIC DESCRIPTION

The metric (SI) system is a logically designed structure of units and reference standards aimed at providing a universal system of measurement. The abbreviation "SI" stands for the French words "Systeme Internationale d'Units" (International System of Units), which is the official name of the metric system. This abbreviation is used throughout the world. The United States is one of the few remaining nations whose citizens do not use the SI system on a day-to-day basis.

Note: Three expressions—"measurement," "reference standard," and "standard unit"—are used throughout this section. The following informal definitions should help to clarify their use:

measurement—the process of determining the size, weight, volume, or some other characteristic of an object, by comparing its observed numerical values to a standard unit; for example, placing a ruler over a sheet of paper to measure its width.

reference standard—the reference standard defines the way the standard unit is to be determined. For example, the metric unit of length could be described as the distance between two etched lines on a certain metal bar; for many years, this was, in fact, the way it was defined.

standard unit—the standard unit is an abstraction; i.e., it is the ideal result of the process described in the reference standard. As a practical matter, we can never construct the perfect inch or the perfect meter on a physical measuring device like a ruler, but the imperfect copies that we can make serve most of our needs quite adequately.

BACKGROUND

From the earliest times, societies have found it necessary to develop units of measure. It was impossible to sell land, livestock, or farm products, or even to cook meals, without the use of some kind of generally accepted standard units.

People resorted to the handiest and quickest reference sources to create such units. If a farmer used a pail of a particular size to sell milk in, that pail might become a standard, with a generally recognized monetary value. A particular cutting of a fabric might be found to be a convenient size to work with; it would become a "bolt," a new standard unit.

Our "inch-pound" system, like all nonmetric general-use systems, grew fitfully and erratically in response to the day-to-day needs of artisans, housekeepers, mechanics, and merchants. The units employed were arbitrary and imprecise, often based on what were felt to be average dimensions of parts of the human anatomy, e.g., the width of the thumb, the length of the foot, the span of the outstretched arm, and the length of a stride.

Throughout the seventeenth and eighteenth centuries, French scientists and philosophers began to question the use of the existing patchwork quilt of units and standards, which differed from country to country and even from district to district within the same country. Commercial and scientific exchanges were becoming more and more important and were made unnecessarily difficult by this disorderly measurement structure. Would it not be better to start from scratch and build a logical and uniform system that could be understood and used everywhere? They decided to make the effort.

The many transformations taking place in France at the time of the French Revolution (1789-1799) helped to open the door to serious consideration of a new system. The actual development of such a system owed its success, in no small measure, to the presence of some of the greatest thinkers of that or any other age. Joseph Louis Lagrange, Pierre Simon de Laplace, Gaspard Monge, Antoine Laurent Lavoisier, and Jean Charles Borda were among the most eminent of this distinguished group.

The benefits of a logically designed measurement structure were quite apparent to them. They formed a committee and focused first on the most important and easily measured characteristic of most objects—the dimension of length. It would be a daunting task to try to list even a portion of the bewildering array of length units that were in use in the countries and territories of the premetric eighteenth century.

The committee members wanted to base their new unit on some fundamental and unchanging physical constant. The length unit should also be of such a size as to maximize its usefulness both upward (into very long lengths) and downward (into very short lengths). It had already been decided that multiples and subdivisions of the basic units would be based on the decimal system, although there had been strong advocacy by some members for a base 12 system.

Almost a century before all of this, a churchman from Lyons, Abbe Gabriel Mouton, had suggested a unit of length based on one minute of arc on a great circle of the Earth. This concept provided the germ of an idea that led the committee to its choice of the fundamental unit.

It defined the fundamental unit of length, the meter, as one ten-millionth of the distance from the North Pole to the equator, along the meridian of the earth passing through Dunkirk, France, and Barcelona, Spain. (A meridian is an imaginary line drawn through the poles and some reference point around the earth.)

Now the task at hand was to actually make the necessary measurement to determine this length. Two surveyors (they were really astronomers) were assigned to carry out the job in 1792. In 1795, before the survey was complete, the general recommendations of the committee on measurement were provisionally adopted at a meeting in Paris.

Although an English representative had attended the meeting, official English reaction was that the system was impractical. The United States was also invited to the meetings but did not send a representative.

Four years later, the two surveyors completed their task. They had actually measured about 1/9 of the selected length of the meridian, which, with the knowledge already available and additional computation, proved to be adequate for their purposes. Their seven-year assignment had included many adventures and misadventures (such as being temporarily jailed as spies), but their measurement (as verified by modern methods) proved to be quite accurate. (They were off by about three hundreds of a percent.)

The metric system was now a reality and was officially adopted by the French government in 1799.

But the cultural hold that the traditional measuring units had on people was not easily overcome. Not only did the new system fail to spread to other countries; public pressure in France, itself, forced the government to allow a return to the use of its old system, 19 years after the original metric law was passed. However, in 1837, the French metric law was reimposed, and it has remained in effect ever since.

The Internationalization of the Metric System

Slowly but surely, after its adoption by the French government, the merits of the new system became apparent, and interest began to develop in other countries.

In 1875, an international meeting was held in Paris, and 17 nations, including the United States, signed a treaty (the Treaty of the Meter), which granted formal approval to the new system. The United States was able to sign the treaty, because Congress had enacted legislation in 1866 making it lawful (but not mandatory) to use the metric system of measurement. Since the signing of the 1875 treaty, the metric system has spread throughout the world (including space), and only a few areas of resistance to its use remain.

One of those areas is the United States, where the inch-pound system remains in everyday use. In the inch-pound system, there is no logical relationship between units, e.g., there are 12 inches in a foot, 3 feet in a yard, 1760 yards (or 5280 feet) in a mile, 16 ounces in a pound, and 2000 pounds in a ton.

It is tedious and time consuming to do calculations in the inch-pound system, and it's difficult to remember such arbitrary unit relationships; but we tend to accept it as we accept other inevitable things we've grown up with, such as the quirks of our language (which often provides us with three different spellings of the same sound) and the vagaries of the weather.

While we have to accept the weather as it is, and while we can cope with language quirks, it's very doubtful that we can continue to operate with an antiquated measurement system when the rest of the world, with whom we must trade, is using a far more efficient one.

THE STRUCTURE OF THE METRIC SYSTEM

The metric, or SI, system has seven fundamental units:

The **meter** for length
The **kilogram** for mass
The **second** for time
The **ampere** for electric current
The **kelvin** for temperature
The **candela** for luminous intensity
The **mole** for amount of substance

There are, in addition, two supplementary units (the radian and steradian) and a large number of derived units.

The basic goal of the creators of the SI system was to determine the most fundamental measurable characteristics of objects and to assign standard units only to those characteristics. (As we see above, the most recent decision is for seven such characteristics.)

The units would be based on enduring reference standards that could be accurately reproduced or copied. (For example, the meter is now defined as the distance that light travels in a vacuum at a tiny fraction of a second.) All other related measurements would then be derived from the fundamental units. For example, we can express the size of the area of a field, not by creating a new unit (such as the acre) but simply by multiplying the length by the width and giving the result in square meters. Volume can be expressed by using the cube of the meter. (The liter which is often used for volume is actually a cubic decimeter or one-tenth of a cubic meter.)

All of the fundamental units (with one exception) are now based on reference standards that can be verified anywhere in the world by any laboratory or testing agency that has the necessary equipment and instrumentation. The one exception is the kilogram. (The kilogram is also the only fundamental unit that carries a prefix.) The kilogram is defined as the mass of a platinum-iridium bar maintained in Sevres, France, in a controlled environment.

It should be pointed out that mass and weight, though often used interchangeably, are not the same thing. Mass can be defined as the quantity of matter in an object. Thus, it is an unchanging characteristic wherever the object is located. Weight, on the other hand, is properly considered a force, because it results from the gravitational pull on an object. When astronauts visited the Moon in the 1960s and 1970s, their mass stayed essentially the same, but their weight was considerably reduced due to the low gravitational pull of the Moon.

We can, however, safely say that the weight of an object is proportional to its mass. Also, since most of us will never leave the earth's gravitational field, we can consider our weight and mass, for all practical purposes, to have the same values.

The unit of volume, or capacity, in the SI system is the cubic decimeter. However, this term has never really caught on, and most people use the word "liter," which identifies the same volume.

The official SI temperature scale is the Kelvin scale. While the size of its unit degree is the same as that of the Celsius degree (formerly called the Centigrade degree), its numerical values differ. The Kelvin zero is defined as that point where there is no heat whatever, i.e., where all energy is absent. At 273.16 kelvins, the zero point of the Celsius scale is reached. Thus, the Kelvin scale has no negative values. At present, the Kelvin scale is largely confined to scientific use, and the Celsius scale is the accepted mode for everyday use almost everywhere in the world.

The SI unit of time is the second, and the United States is currently using it in conformity with the rest of the world.

Very large and very small dimensions can be dealt with by using a series of prefixes which rise or descend in multiples of ten. Thus, on the descending scale there is a decimeter (1/10 of a meter), a centimeter (1/100 of a meter), and a millimeter (1/1000 of a meter). The scale continues further, but these are the dimensions likely to be encountered in everyday situations. On the ascending side of the scale, there is the dekameter (10 meters), the hectometer (100 meters), and the kilometer (1000 meters). The words "dekameter" and "hectometer" aren't heard very often; people find it easier to say 10 meters or 100 meters. The word, kilometer, of course, is very familiar. The point of these prefixes (which may be used with any fundamental unit) is that they enable us to have very large and very small units without inventing new names and new relationships.

The calculating advantages of the metric system are clear. Because all units are expressed in multiples of 10, they may be added, multiplied, and divided with remarkable ease. In contrast, the irregular relationships of the inch-pound system make for tedious and confusing calculations.

Using the SI System in Everyday Life

The units that we are most likely to encounter, as the United States converts to SI (or if we pay a visit to virtually any foreign country), are the units of length, the meter and kilometer; the unit for weight, the kilogram; the unit for temperature, the Celsius degree;

the unit for volume, the liter; and the time unit, the second (with which we are perfectly familiar). Ordinarily, we would be seeing signs and getting directions in SI units and trying to visualize them in inch-pound units, rather than the other way around. The U.S. Department of Commerce suggests the following approximate quick-conversions (some informal visualization aids have also been provided):

D.C. approx. conv.	Visualization aids
0.04 x number of millimeters = inches	(a millimeter is little more than 1/32 of an inch)
0.4 x number of centimeters = inches	(there are about two and a half centimeters in an inch)
3.3 x number of meters = feet	(there are little more than 3 feet in a meter)
0.6 x number of kilometers = miles	(a kilometer is a little more than half a mile)
1.06 x number of liters = quarts	(a liter is almost equal to a quart)
2.2 x number of kilograms = pounds	(a kilogram is slightly more than two pounds)
9/5 x number of Celsius degrees + 32 = Farenheit degrees	(In the range of ordinary weather conditions, doubling the number of Celsius degrees and adding 30 will get you close enough to the Fahrenheit equivalent to keep you going until you find a calculator)

THE FUTURE

While the SI system represents an enormous advance for science and industry and is a key factor in the modern high-tech world, it is still evolving. (For example, there is, as yet, no SI unit for sound intensity.) An international committee meets at regular intervals to consider suggestions and changes.

As indicated above, the units of measure that a society develops tend to become deeply ingrained in the cultural life of that society. In the United States, the Indianapolis 500-mile speedway race, the "quarter-pounder" hamburger, the miles-per-gallon performance of our cars, and the 100-degree days of July and August are familiar topics of conversation in virtually every household in the nation.

It happens, however, that every unit of measure used in the above examples has either been abandoned by the rest of the world or was never used at all. The rest of the world is using the SI system. Although Congress has called for adoption of the metric system as our primary measuring procedure, there appears to be little enthusiasm for this policy on the part of the average citizen. The old units are familiar and comfortable. (The United States is joined, incidentally, in its resistance to the use of the SI system by Myanmar [Burma], South Yemen, and Tonga.)

In the Trade and Competitiveness Act of 1988, Congress directed that starting November 30, 1992, every federal agency must purchase only products manufactured in SI units, unless they are unavailable in this form. But by the latter 1990s, there was still resistance, not only from the public, but also from those in business and the government not wanting to spend money for such a change. Apparently, the metric system in the United States still has a long way to go.

It is clear that the conversion to SI units in the United States is both an industrial and a social change. (Science has long since adopted the SI system.) The industrial change is well underway (very late by contrast with the rest of the world but finally picking up steam), as is the change to SI for the United States government. The social change is proceeding with much less conviction.

It has been pointed out that many people, particularly older people, who depend on such things as early morning radio temperature readings to know how to dress for the weather, could be terribly confused by Celsius numbers. There might also be quite a bit of confusion for a while when filling the gas tank. But our neighbors to the north in Canada have handled these changes without earthshaking consequences. In some situations, dual readings could be provided, for a while, to smooth the way. The social change is necessary, because were it not to take place it would saddle manufacturers with the need to supply some products in the two different measurement systems.

The benefits of full conversion to the metric system in the educational system will be substantial. Students still operating within the archaic structure of the inch-pound system are wasting hours of precious class time on tedious calculations that are only necessary because of the inadequacies of that system. Although many classrooms now teach both systems of measurement, there is still a strong opposition to the change.

There is little doubt that into the next century, the entire world, including the United States, will have to embrace the new global economy, which means everyone will be making measurements in the units of the SI system.

FURTHER READING

Books

Metric in Minutes: The Comprehensive Resource for Learning and Teaching the Metric System. Dennis R. Brownridge. New York: Professional Publications, 1994.

A comprehensive introduction to the SI system.

Metric Units and Conversion Charts: A Metrication Handbook for Engineers, Technologists, and Scientists. Theodore Wildi. New York: IEEE, 1995.

A scholarly, but very readable study of the SI system.

SI: The International System of Units. Robert A. Nelson. Stony Brook, NY: American Association of Physics Teachers, 1982.

Contains a brief history and commentary.

Articles

"Guide for Metric Practice." R.A. Nelson. *Physics Today*, August 1996, B615-B616.

A look at the metric system.

"No, Let's Keep America American." N. Brunt. *Popular Mechanics*, September 1996, 46-47.

The continuing arguments to keep America away from the metric system.

"Yes, America Needs to be Metrified." R. Bonner. *Popular Mechanics*, September 1996, 46-47.

The other side of the argument to get America to turn to the metric system.

SECTION THREE
Brief Descriptions of Selected High-Tech Items

This section discusses a group of high-tech items that form part of the major systems and devices described earlier, or operate as stand-alone units. Some are commonly used in the home or in the office. Others are seen in stores and libraries. Some items (transistors and diodes, for example) are present in almost every electronic device that we buy. The purpose of this section is to provide a brief overview of each of the selected items.

Selected High-Tech Items

BAR CODING

Although we are used to seeing bar code devices in supermarkets (where they have been in use for more than 20 years), they are now also appearing in libraries, factories, warehouses, department stores, military installations, auto parts distributors, and health organizations. We see them on newspapers, magazines, mailing envelopes, beer cans, milk cartons, and soft drink bottles.

Bar coding is a form of optical storage (+). Its principle of operation is basically the same as that of compact discs, CD-ROMs (+), and optical computer memory. That principle is simple in concept: you print (or otherwise transfer) a message, coded in marks or symbols, onto a surface (such as that of a disc, or a paper label, or even a soft drink can); later, when you need to decode the message, you shine a tiny laser beam (+) or LED (+) light onto the surface and "read" the marks by monitoring the reflected light. (The pattern of the reflected light will vary, depending on the mark it is reflecting from.) The varying reflections are then converted into electric impulses and decoded by a computer.

Bar code devices differ in one important respect from most other optical memory devices: they are not able to store nearly as much information on a given area of surface as a compact disc or a CD-ROM. While a unit of data on a CD-ROM is so tiny that it isn't even visible to the unaided eye, the same quantity of data in a bar code system will be quite visible; in fact, as visible as any ordinary printed alphanumeric character.

This "inefficient" information storage in bar coding is not an accident; it's inherent to the process, because bar coding is an "open system"; i.e., its coded messages are printed in a variety of different printing plants on a variety of different surfaces. This diversity can bring considerable variation in the quality of the printing. In addition, the messages are read under "real-world" conditions by a host of different instruments. They are also used in many different environments, ranging from dim warehouses to brightly lit supermarkets.

Much bar code reading is done with a "wand," i.e., a hand-held device that emits a tiny beam of laser (or LED) light and then monitors the reflections, as described above. The wand is not always held the same distance from the marked surface, nor is it always passed over the surface at the same speed. It is often passed from right to left, as well as from left to right. (The Universal Product Code, used in most supermarkets, can be read in either direction.)

At the supermarket, the bar code will probably be read by brushing the commodity unit over a stationary reader built right into the counter. The angle at which the code strikes the reader may vary, as well as the speed at which the code passes the detector. In addition, the marks on a can or bottle may be hard to read because of curvature and the differences in natural reflectivity of these surfaces. The codes on any surface may be impossible to read if ink spread from a poor printing job has made the edges of the bars indistinct. It is clear, then, that bar code signals must be extremely "emphatic" and "powerful" to avoid being distorted or misread in the "noisy" and changing environments in which they are produced and used.

In contrast, most compact discs and CD-ROMS are "closed" systems in which the manufacturer completely controls the quality and specifications of the discs. When the discs are being "read," they are enclosed within a tightly sealed compartment where light intensity, reading distance, speed, and direction are all rigidly specified and regulated. This allows them to store many more magnitudes of information than can bar codes. However, since bar codes generally carry relatively brief, uncomplicated messages, their limited storage capacity is usually adequate.

Probably the most prevalent code system is the one used in supermarkets and some department stores; it is known as the Universal Product Code. The bars in this code represent numbers that identify the manufacturer of the item, as well as the size, and type of product. It's interesting to note, however, that the code does not include the price of the item. When the identifying code is read, the information travels over the line to the computer database, which checks it against its price list and sends the price right back to the cash register, all in a fraction of a second. This set-up allows prices to be changed without having to mark individual shelf items. Some stores even provide bar code readers throughout their stores in lieu of price tags or prices printed on the shelves.

CAD (COMPUTER-AIDED DESIGN)

Computer-aided design (CAD) programs are expert systems that help technical designers create new products or improve existing products.

CAD systems are particularly important (in fact, indispensable) in the design of new integrated circuits (chips [+]). At times, the CAD process becomes a kind of dialogue between the designers and the computer, via the keyboard, display screen, and printer. By querying the program, the designer can find out, almost instantly, the answers to such questions as: If I change this value, what happens to the rest of the configuration?

CAD programs contain whole libraries of component designs. They can also generate detailed instructions for automated manufacturing equipment as well as parts lists, blueprints, and specifications.

In effect, the CAD program takes the designer's creative ideas and fits them into the realities of product dimensions, material characteristics, and time constraints. It speeds up the entire process, reduces human error, and permits a greater number of designs to be considered.

CAM (COMPUTER-AIDED MANUFACTURE)

Computer-aided manufacture programs guide automated equipment through the manufacturing process. The computer may employ sensors (+) to monitor the process, issuing changes of direction and corrections based on its analysis of the reported data. Plants that have reached a high level of automation are able to perform under the direction of such programs with very little human intervention.

It is possible to envision the time when the plant computer will be able to take the (computer-generated) design of a new product, assemble required materials and supplies, allocate resources, schedule machines, and proceed to manufacture the product with little or no help from people. Some pilot plants are already attempting this level of automation with relatively simple products such as planar parts (flat objects) to be used in assemblies. Sheet metal, paperboard, plywood, and fabric have also been processed in automated plants.

CCDs (CHARGED-COUPLED DEVICES)

Charged-coupled devices are image sensors that change light into electronic signals. The actual device consists of a layer of silicon and a layer of insulating silicon dioxide, both above a base of metal electrodes. The CCD gathers photons (light) that strike the silicon. Electrons are released where the photons strike the device, essentially creating an electronic version of the picture; the brightest areas collect the largest amounts of electrons as the most photons strike those areas.

This light-sensitive technology makes it easy to translate certain images from the electric signals to computers, to store them on computer disks or hard drives, or to transfer them over electric or telephone lines. CCDs are found in almost every device that transfers images, including electronic cameras, television cameras, and telescopes. Video images (the ones seen on your television or computer) are produced with CCDs: For example, and simply put, inside a video camera that produces a TV picture, the reds, greens, and blues of the image are separated, their light falling on the CCDs, which, in turn, create a certain pattern of electric charges. The video signals that go to your television set are actually these charges being released line by line. (Essentially, when the signals reach your television, the video signals are transmitted to three electron guns. The three resulting beams are guided by magnets around the collar of the tube, and without getting into too much detail, are deflected from left to right and top to bottom, "painting" lines on the screen—at around 1/30th of a second.)

In the late 1970s, astronomers began to replace film cameras on telescopes with CCDs. After pointing the telescope at the desired area, the CCD collectors capture the light and translate it into an electric signal. The signal is then transferred to a computer (CPU) to be translated immediately onto a computer monitor. The light sensors collect more light than film does, allowing astronomers to see some of the fainter clusters, galaxies, and stars in the universe.

Another use for CCDs has recently developed: fingerprint identification devices for your personal computer, which uses CCDs to scan the user's fingerprints. This type of device is mainly for security reasons and to restrict computer access.

CERAMICS

Ceramics is an old technology that has found important new applications in the high-tech world. Because of their resistance to high temperatures and hostile environments, ceramic products are used in nuclear reactors, in furnaces, in the linings of jet engines and rocket nosecones, and as protective tiles on the outside of rockets. They are also used in electronic components as insulators and in magnets and microphones.

A ceramic is made by firing or baking a nonmetallic mineral (clay is the common raw material) under controlled conditions. One problem with ceramic materials is that they tend to be brittle; however, recent improvements have diminished this problem significantly.

Ceramics are particularly effective in transducers (devices for converting acoustical energy, or other forms of mechanical energy, into electric current). They are also widely used in a broad assortment of critical electronic components and in permanent magnets.

Recent advances in superconductivity (+) will create important new applications for ceramic materials. Various ceramic oxides have been developed that work as superconductors at temperatures close to those maintained in ordinary home freezers. If these materials can be transferred successfully from the research laboratory into commercial use, they will bring revolutionary changes to everyday living. (The most immediate impact of successful high-temperature superconductivity for the average citizen would be much cheaper electricity, because energy would be processed and distributed much more efficiently. Other effects of this new technology are discussed in greater detail in the section on superconductivity.)

CIRCUITS

To run current flow through various devices, circuits are needed. A circuit consists of the wires, the signal lines, and the electric components that carry a current or any collection of componets (or an infrastructure) that allow a current to be turned on or off for a specific purpose. (For example, a circuit to turn on or off a light would differ greatly from a circuit for a scanner.). A simple example is a lamp that has a switch in the circuit that turns it on and off. When the switch is on, the current is flowing (closed circuit); when the switch is off, the current is broken (a gap separates the connecting wires creating an open circuit). Circuits are used in every electronic device. For example, circuits allow a keyboard to be connected to the rest of a calculator. As a key is depressed the circuit is open; it

sends out an electric charge, which is translated as a signal representing a number or operation. Circuits are also used in electric motors, household appliances, long-distance power lines, and other devices. Circuit breakers are protective devices that let only a certain amount of electric current through to a device or to a household (from the electric lines). If the electric current is too high because of a lightning strike or power surge, the circuit breaker automatically trips open the circuit and cuts off the electric current.

COMMUNICATIONS SATELLITES

Earth-orbiting satellites have become the primary factor in long-distance communications around the world. (A primacy that is seriously threatened, in the opinion of some observers, by the rapid growth of fiber optics [+] systems.)

In contrast to Earth-based relay systems, whose signals are blocked by natural obstructions and hampered by the Earth's curvature, satellites, properly positioned, can transmit from one point on Earth to any other.

Modulated (signal-carrying) electromagnetic waves (+) are beamed up from transmitting stations on the Earth's surface to an orbiting satellite. The signal waves may be carrying TV programs, telephone calls, or computer data. The satellite either bounces the signals down directly to users or to an Earth-based receiving station, or transmits them to another satellite closer to the Earth station. The receiving stations on Earth process the signals and distribute them to users.

Most satellites amplify (and otherwise process) signals before they are retransmitted to Earth. Power for their electronic gear is usually generated from solar cells.

The majority of communications satellites are traveling in "geostationary" or "geosynchronous" orbits. These terms mean that their speed and orbiting paths are so calculated that they stay above a fixed spot on Earth. If you were able to look up and see them, they would appear to be standing still. This is why receiving antennas on building roofs and in people's backyards can be aimed at a particular spot in the sky and left in the same position indefinitely.

CRT (CATHODE RAY TUBE)

A cathode ray tube is a vacuum tube that generates and focuses a stream of electrons onto a phosphor-coated screen. The screen reacts to the stream of electrons hitting it by fluorescing (emitting light) and creating visible images on its external (display) side.

The electron stream is deflected and modulated, i.e., varied in intensity, to carry information such as a TV picture or a computer readout.

The CRT is the basic display unit of the television and computer industries. The next step in such units is the high-definition television. As previously explained, electron streams are channeled line by line on to the screen. The more lines per screen, the better the resolution. The high-resolution televisions will produce 1,125 lines for greater detail, almost double the conventional television with 525 (in Europe, 625) lines.

GPS (GLOBAL POSITIONING SYSTEM)

The global positioning system is an accurate way of determining a person's position on the surface of the Earth. Once used only by the United States Department of Defense (and they still have a better GPS system than is commercially available), the GPS uses a collection of 24 Navstar satellites, each tuned with an onboard atomic clock. In general terms the system works as follows: A person on the ground, in the ocean, or in the air sends a signal from a hand-held device to the satellite system; the atomic time on the satellite is then compared to the surface device's inbuilt clock and the resulting signal is returned. Because the speed of radio waves are known, and the GPS unit "knows" the exact position of the transmitting satellite, an accurate position can be determined usually in the form of a map or latitude and longitude readings, depending on the device. GPS signals are better than most navigational devices, because they can be used in any weather, season, or location.

HOLOGRAPHY

Holography is the process of recording, and later reconstructing, three-dimensional visual images of physical objects.

The process uses sophisticated optics to split a laser beam into two separate parts. One beam is directed at an object, from which it is reflected toward a photographic plate. The other beam (called the reference beam) is aimed directly at the plate. The two beams meet at the plate and form an interference pattern that is recorded on the plate.

When the original laser beam (the reference wave) is later passed through the "hologram" (the photographic plate with the interference pattern on it), an accurate and detailed three-dimensional image of the recorded object becomes visible.

Important applications of holography are anticipated in computer data storage and medical diagnostic techniques. Since a hologram records three-dimensional information on a flat surface, its information storage potential is much higher than that of conventional optical storage.

A group of American researchers at Northwestern University is working on a holographic technique that would allow some imaging of internal body tissues using visible light, thus avoiding the ionizing radiation (+) of X-rays. This process has successfully imaged through two and a half inches of breast tissue, indicating some potential for other medical applications.

INTEGRATED OPTICS (IO)

An integrated optics device does essentially what an integrated circuit (+) does but uses light, instead of electric current, as the essential medium. If light-based computers become, as expected, the next important advance in computer technology, integrated optics will play a fundamental role.

The optical equivalents of switches and other components are used in integrated optics. The advantages of IO include higher switching speed, little or no generation of heat, immunity to electrical noise, and greater signal-carrying capacity.

Many significant engineering problems are still to be solved, however, before IO-based systems become widespread. For example, converting light signals to sound signals, as in a fiber optics (+) telephone system, still requires elaborate intermediate electronic devices.

While the chief impact of integrated optics will be felt most strongly in the field of computers, it is reasonable to expect that as light, in place of electric current, begins to appear in consumer devices, we will see smaller, cooler-running, more efficient household appliances.

LCD (LIQUID CRYSTAL DISPLAY)

A liquid crystal display is a device that forms visible, but nonradiant, characters or images on a screen.

In other words, the LCD display is visible to the viewer only through reflected light. In darkness, it's invisible. (It's like writing on a blackboard. You can't see it if the room lights are off).

Its great advantage is that, since it generates no light, it uses very little power. LCD display watches can operate for years on the same tiny battery. Laptop computers are practical only because they use the LCD principle.

An LCD device functions by storing a clear liquid crystal material between two sheets of glass that have been coated with a transparent, electrically conductive material to form an electrode system. The coating on the display side is etched into character-forming segments that are wired to external connections. When a voltage is triggered by a keyboard input, or some other source, the applied voltage disturbs the orderly molecular arrangement of the liquid crystal, causing it to darken sufficiently to form legible characters through the appropriate etched segments.

The quality of LCD images has shown steady improvement over the years, and, because they do not require the physical depth of a CRT display, they have begun to appear in small portable TV sets. Most current LCD television, and some laptop or smaller computer monitors, use an "active matrix" (or thin film transistor [TFT]) design. The screen is "active" because each tiny, independent transistor (actually, from one to four transitors) turns each pixel current on or off, building up the image on the screen. The biggest performance difference between the active matrix LCD and standard LCD (also called "passive") is speed: The transistors mean that current that triggers the pixel illumination can be smaller—thus, turned on and off faster. Plus, it's faster for the user: On a passive screen, when the user uses the mouse to move the cursor, it takes time for the display to "catch up" on an active screen. The display is fast enough to see the cursor move at all times.

MICROSCOPES

Microscopes have changed a great deal since Hans Lippershey first invented the compound microscope in 1609. Today's smaller microscopes, such as the binocular and monocular microscopes, still use the same principle: Glass lenses are used to magnify the images of objects, but unlike the crude glass beads used as lenses long ago, modern microscopes use a number of large glass elements, allowing researchers to see more detail at better resolution. Cameras, video cameras, and computers have also been fitted to modern microscopes, allowing for images to be captured on film or a computer screen, to be used for various scientific studies.

Other types of microscopes, however, allow scientists to study objects in much more detail. Two are based on the emission of electrons: Transmission electron microscopes (TEM) use thinly cut samples that electrons can pass through, giving an image of a structure (based on whether the electrons pass through the sample or not, depending on the density of the sample), such as mineral grain boundaries or defects in metals. Magnification can reach a hundred thousand times normal size. Scanning electron microscopes (SEM) are based on the same technique, but the entire surface of the object can be imaged to see the details due to the reflection (or scattering) of electrons, which are analyzed by special detectors. Perhaps the most powerful microscopes are the scanning tunneling microscope (STM) and the atomic force microscope, which are usually used to examine objects at the atomic level. Simply put, the contours of a surface can be mapped by determining the variation between the tip height (of a specially designed tip) and the sample surface—based on the amount of current flowing from the tip and surface. Most atomic force microscopes also use a special tip that actually touches the surface sample; the measurements of the tip's up-and-down movements creates an "image" of the atomic surface of the sample.

MICROWAVE OVEN

The segment of the electromagnetic spectrum (+) between wavelengths 0.3 and 30 cm (approximately) is identified as the microwave region of the spectrum. The microwave region has proved to be a fertile source for many modern technical devices. Shortwave radio, radar (+), and wireless telephones utilize energy in this neighborhood of the spectrum.

One familiar high-tech device that utilizes electromagnetic radiation in this region is the microwave oven, which has changed the way many people in industrialized societies prepare their food for the table. Microwave ovens can now be found in more than 75 percent of American homes.

A conventional oven heats food by first heating the air in the oven. The hot air then warms the surface of whatever food is being prepared, and the heat gradually spreads from the surface to the food's interior, warming (in effect) from the outside in. This process is usually a slow one.

The microwave oven, on the other hand, heats food by using electromagnetic waves that instantaneously penetrate into all layers of the food and proceed to interact with and excite the food and water molecules (molecules are bonded groups of atoms) that make up the substance. As we have noted earlier, the temperature of a substance is actually a measure of the activity of its molecules; if they're active, it's hot; if they're inactive, it's cool (relatively speaking). The quick-acting, energizing effect of the microwaves on the molecules, particularly the water molecules, gives us evenly warmed food in a very short time. In principle, the heating action of the microwaves on the food is similar to the way electric current heats the filament of a lamp. In practice, the microwave oven may sometimes produce unevenly or inadequately heated food due to excessively short cooking times recommended by optimistic manufacturers for particular kinds of food.

The microwave oven generates its electromagnetic waves through the use of a "magnetron." (A magnetron is an electronic tube that produces high-frequency microwave oscillations; it's the same device used in radar.)

OPTICAL CHARACTER RECOGNITION (OCR)

Optical character recognition (OCR) is a close relative of bar coding (+). The primary difference is that the characters read by OCRs can also be read by people. The basic operation of OCR devices is to scan a printed surface with a point of light, monitor the reflections, and compare the character scanned with a reference character in computer memory. Within a controlled system, i.e., a system in which the characters are of a particular style and size, these systems are very successful. They have been in use since 1955 and have been particularly effective in banking applications.

It is much more difficult for these devices to "read" uncontrolled printing and handwriting. Rapid progress is being made, however, and some of the more sophisticated devices are attempting to use "context," i.e., the surrounding letters and words, to determine what letter a difficult-to-read character is intended to be.

This technology is of particular interest to the postal service because it could eliminate the labor-intensive keyboarding required to enter zip codes into automatic sorting equipment. However, using OCR devices presents difficulties because the post office not only gets a vast assortment of printing styles and handwritings, but there is also wide variation in the position of addresses on envelopes.

Many applications, other than those of the post office, require massive data entry (government agencies and online databases [+], for example), and the successful use of OCR devices would mean enormous savings.

PLANETARY SPACECRAFT

Somewhat similar to communications satellites, planetary spacecraft are used to explore the various planets and their satellites in the solar system. Because of the great distance involved in this interplanetary exploration, they are usually powered differently than communications satellites. Most short-distance communications satellites carry conventional power onboard, including batteries and solar panels, as they are usually within close proximity to the Sun's rays that give power to the solar arrays. Long-distance craft are usually powered with nuclear power, as they are too far from the Sun to make solar arrays useful.

Both long- and short-distance craft are controlled by beaming modulated (signal-carrying) electromagnetic waves (+) from transmitting stations on Earth to the craft, usually carrying data that will tell the craft what maneuvers to make and functions to perform. Most planetary spacecraft carry multiple sensors onboard to analyze a planetary body. Resulting data (in the form of digital images [+] and other sensory information), also in the form of signal waves, are sent back to the Earth; fainter signals from long-distance craft are often picked up by large collections of radio telescopes. The Earth receiving stations then process the signals and distribute them to users.

The majority of the planetary spacecraft travel past the planetary body, collecting data in a flyby, in which the craft swings past the object without stopping, such as the Voyagers 1 and 2 spacecraft flybys of the major outer planets. Other craft go into orbit around a planetary body to take detailed images of the planet. The Viking orbiter to Mars, the Magellan spacecraft to Venus and the Galileo at Jupiter are only three of the many successful crafts that have orbited and imaged other worlds. Still other planetary spacecraft have landed on a planet, including the Viking lander on Mars, and the Mars Pathfinder, which in 1997, also sent out a rover, the Sojourner, to explore the area around the lander. Other future planetary crafts will image solar system members in the future, including the Mars Global Surveyor to Mars; Cassini/Huygens to Saturn; and the Lunar Prospector to the Moon.

RADAR

Radar accuracy, resolution, and use has advanced in the last decade, and in its various modifications is used extensively in airports, on aircraft and ships, by police forces (mostly to catch speeders), on Earth-observing satellites, to detail the ocean floor, and for weather forecasting. In general terms, radar works by bouncings a signal off a far-away object. (Radar is an acronym for *r*adio *d*etecting *a*nd *r*anging.) The radar picks up the reflected returned signals with a special receiver and translates them into useful data.

For example, radar in airports use primary and secondary radar usually originating from the control tower: The primary radar uses microwave energy beamed from a rotating antenna that searches the sky in all directions. As a plane passes, the waves bounce off it, are then collected again by the antenna, and sent back to a radar tracking screen. Secondary radar, more advanced and usually located at a nearby building, bounces signals off the plane, which in turn sends back more detailed information about the airplane, such as its four-figure call sign and its altitude. Both radar data are then combined—and both are necessary to allow air traffic controllers to control the airspace around an airport.

The use of Doppler radar in weather forecasting has become valuable in detecting normal weather features and hazardous outbreaks of bad weather (Doppler radar is also used by police to detect speeders). Doppler radar is based on the Doppler effect: As a wave sent out by the radar hits an object approaching the radar, the wave is reflected at a higher frequency than the original wave sent out; if the object is moving away, it will be reflected at a lower frequency. The larger the frequency difference between the outgoing and incoming (reflected) waves, the faster the speed of the object. Certain details in the data—especially tracking small scale changes within a storm with the Doppler radar—are used to determine weather conditions that have the potential to form tornadoes, important in reporting tornado watches and warnings to the public.

SEMICONDUCTORS

Most metals are good conductors of electricity; most glass and porcelain materials are not. Metals conduct well because they contain many free electrons. In glass and porcelain insulators, electrons are tightly bound to their atoms and cannot conduct current. (See "Electromagnetic Spectrum.")

Semiconductors fall somewhere in the range between conductors and insulators. In their pure state, at room temperature, they can conduct only slightly because they have only a few free electrons. The most common semiconductors are silicon and germanium.

But semiconductors possess properties that set them apart from other materials and make them vitally important to the new technology. Their most important characteristic is their versatility. For example, they can be made to give off light when an electric current is applied, or, conversely, to convert light into electric current. Their level of conductivity can be raised or lowered significantly. Moreover, conductivity can be maintained at varying levels in different local areas within a single tiny square of semiconductor material. Thus, we can have a tiny object that is a strong conductor in many local areas and an insulator in others. In addition, it is possible to change an insulating area to a conducting area, or vice versa, in a fraction of a second.

As a result designers can build electric devices like switches right in the semiconductor material itself. In fact, it would be possible to build entire miniature versions of household wiring circuits within a single tiny block of semiconductor material. This ability of semiconductors to change their state and to maintain conductive and nonconductive areas or spots, as needed by circuit requirements, has made them the fundamental cornerstone of electronics and has resulted in the creation of a new branch of science called solid-state physics.

Changing the Level of Conductivity of a Semiconductor

In their normal state, semiconductor atoms share electrons (+) with their neighbors to form a tight structure that depends on the proper ratio of electrons to nuclei. Because of this balanced structure, semiconductor materials have few free electrons and cannot conduct current to any great extent. However, this orderly structure can be disrupted by introducing minute quantities of "impurities" into the material. (This procedure is referred to as "doping.") Some "impurity" atoms will bond with some of the semiconductor atoms and will free up electrons. The "doped" part of the semiconductor will then be capable of conducting current. In the case of silicon, this change can be brought about by injecting tiny quantities of phosphorous onto the silicon. The segment so treated will now have conducting electrons within it and is referred to as an "n" (for negative) type carrier.

Alternatively, an impurity such as boron can be injected, and its interaction with the silicon atoms will produce a surplus of positive carriers in the treated segment. The positive carriers produced this way are referred to as "holes" (because, in essence, they represent the absence of an electron). A silicon segment doped with boron is referred to as a "p" (for positive) type carrier.

Semiconductor Diodes

A charge is a force that repels like charges and attracts opposite charges. If a p-type carrier is joined to an n-type carrier (during the manufacturing process), and the combined segment is connected to a battery so that the lead from the positive terminal is connected to the "p" side and the negative terminal joined to the "n" side, current will flow through the circuit. (This type of connection is referred to as "forward biasing.")

Current flows because (positively charged) "holes" in the p-region are repelled by the positive terminal and (negatively charged) electrons in the n-region are repelled by the negative terminal. Thus, carriers flow through the pn junction (the area where the two segments join) and all across the combined segment, with the holes now attracted by the negative terminal and the electrons attracted to the positive terminal of the battery.

However, if the positive terminal of the battery is connected to the "n" side of the joined segment, and the negative terminal is connected to "p" side, there will be no current flow. (This is referred to as "reverse biasing.")

Current ceases to flow because the positive terminal strongly attracts the electrons in the n-region, the negative terminal attracts the "holes" in the p-region, and a wide area depleted of electrons and "holes" forms at the pn junction. This area will, effectively, have become an insulator, shutting down the current.

The pn semiconductor is an important electronic device and is referred to as a "diode." (See figure 20.) Since it will permit current flow in only one direction, it can convert alternating current (AC) into direct current (DC), and it has many vital applications in industry and science. (Many devices run on direct current, for example, many types of motors, calculators, and portable radios. When one wants to use an AC source such as a household outlet to run these devices, a diode can be used to convert the alternating current to direct current.)

Light-Emitting Diodes (LED)

A light-emitting diode is a forward-biased semiconductor device that gives off visible light when electric current is passed through it under certain conditions.

Since LEDs are actually radiating light, they can be seen under any conditions, including total darkness. They are long-lived, reliable, and provide very clearly defined images. However, they are not generally used in devices that are totally dependent on batteries, because, unlike LCDs (+), they are significant power consumers.

Photodiodes

A photodiode is a reversed-biased diode that reacts to light by generating an electric current. The amount of current produced is proportional to the amount of light falling on the diode.

Photodiodes are very useful as light-operated switches or counters.

Note: In practice, the term "semiconductor" is used to describe devices made of semiconductor material as well as the material itself.

SENSORS

A sensor is a device that is sensitive to certain conditions in its environment, such as light, temperature, pressure, motion, and the presence of particular substances. Ordinary cameras are sensors, as are spy satel-

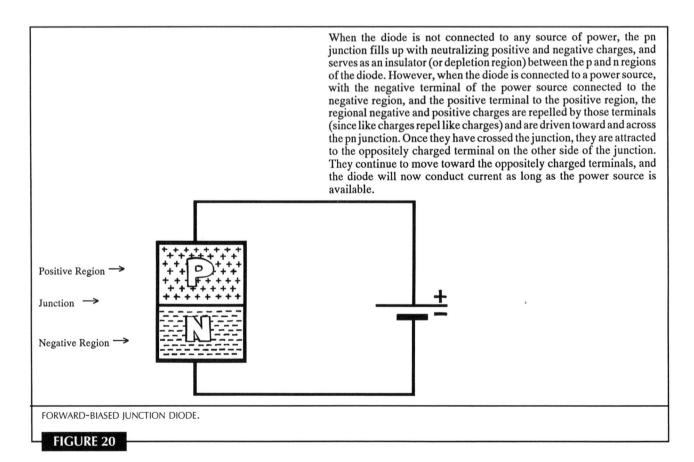

When the diode is not connected to any source of power, the pn junction fills up with neutralizing positive and negative charges, and serves as an insulator (or depletion region) between the p and n regions of the diode. However, when the diode is connected to a power source, with the negative terminal of the power source connected to the negative region, and the positive terminal to the positive region, the regional negative and positive charges are repelled by those terminals (since like charges repel like charges) and are driven toward and across the pn junction. Once they have crossed the junction, they are attracted to the oppositely charged terminal on the other side of the junction. They continue to move toward the oppositely charged terminals, and the diode will now conduct current as long as the power source is available.

Positive Region →

Junction →

Negative Region →

FORWARD-BIASED JUNCTION DIODE.

FIGURE 20

lites, thermostats, smoke detectors, weather vanes, and burglar alarms.

Sensors are particularly helpful when they can be placed in hard-to-reach locations such as outer space, hot furnaces, working engines, deep water, and internal organs of human beings. In such situations, they provide information that can be obtained in no other way.

In addition to their ability to "read" some condition of a particular environment, sensors must be able to report back to "headquarters," so that the information can be utilized. From outer space, the reporting will be by radio; from deep sea waters, it will probably be through fiber optics strands. A host of other reporting methods are used.

Sensors are also particularly helpful in situations where constant surveillance is necessary. In very high-precision manufacturing processes, sensors may be used to provide an immediate warning when some phase of the operation goes out of tolerance. Hospital patients who are critically ill may have sensors attached to them to warn immediately of any change in vital functions.

Sensors have become particularly important in the computer age due to the computer's ability to accept and analyze huge chunks of data with great speed. This speed is important in complex manufacturing operations and in military situations where reaction time is often extremely critical.

The combination of the sensor and the chip (+) is a particularly potent one; the ability of the chip to control a machine, or a physical process, makes possible a high degree of automation.

Silicon, the most familiar semiconductor (+) material, makes a particularly good sensor because of its sensitivity to many environmental conditions; it is also a tough material, quite resistant to hostile surroundings.

Modern automobiles are prolific users of sensors. A typical vehicle may contain as many as eight sensors in its power-train system alone.

TRANSISTORS

The transistor is an important semiconductor (+) device that is used for amplification, switching, and rectification (changing direct current to alternating current).

Transistors are present in almost every electronic device that we buy. In its important role as an amplifier, the transistor takes a very weak signal (as from a distant radio station) and uses it as a guide to regenerate a strengthened (but faithfully reproduced) signal

in a strong current drawn from an external source. This is like a band leader using a weak signal (the movements of a baton) to modulate a very strong signal (the sound of the instruments) as the entire marching band (the external energy source) follows the baton.

If (in the manufacturing process) we join a diode (+) containing a large n-region and a narrow p-region to another large n-segment, and connect the two n-regions to a battery as shown, no current will flow, because a wide depletion region (an insulator) will form at the pn junction between the n-segment connected to the positive terminal of the battery and the p-strip, just as it would in a reverse-biased diode.

However, if we connect the p-region to the positive terminal of a separate battery and allow a weak current to flow, electrons in the forward-biased n-region will be attracted to the p-strip and will sail through the pn junction, across the p-strip, through the second pn junction and into the reverse-biased n-region. They will not linger in the p-strip, because it's so narrow and because they are already coming under the attraction of the strong positive terminal linked to the reverse-biased n-region.

In electronics terminology, the forward-biased n-region is called the "emitter," the narrow p-strip is called the "base," and the reverse-biased n-region is called the "collector." (See figure 21.)

The flow of current from the emitter to the collector can be increased or decreased by varying the strength of the base current. (The base current is the transistor's "baton.") The important point is that the large emitter current is controlled by, and is proportional to, the small base current. By placing a signal on the base current, we can obtain a greatly amplified (but accurately reproduced) version of the signal in the emitter current. Amplification is, in fact, one of the most important functions of the transistor. The transistor can also serve as a switch. Reducing the base current to the point where no emitter current flows, and raising to reactivate emitter current, etc., provides this switching action.

Transistors appear in many other forms, and they are the most important of the semiconductor devices.

FURTHER READING

Books

The Bar Code Book. Roger C. Palmer. New York: Helmers Pub., 1995.
 A sourcebook for bar coding.
Basic Principles of Semiconductors. Irving M. Gottlieb. New York: H.W. Sams, 1996.
 Concise discussion of electrical concepts of semiconductors.

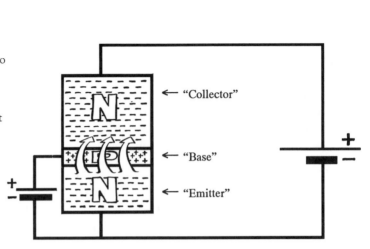

When the narrow p-region (called the "base") is connected to the positive terminal of a second power source, a weak current consisting of "holes" (positive charges) flows into the base. The holes attract electrons (negative charges) from the forward-biased n-region (called the "emitter") which flow into the p-strip and then into the reverse-biased region (called the "collector"). The flow of electrons into the collector is controlled ("modulated") by the base current, and a signal on the base current can be reproduced in the much stronger current flowing from the emitter.

NPN JUNCTION TRANSISTOR.

FIGURE 21

The CAD/CAM Handbook. Carl MacHover. New York: McGraw Hill Text, 1996.

A comprehensive guide to the hardware, software, and applications technology available for CAD/CAM.

Handbook of Modern Sensors: Physics, Designs, and Applications. Jacob Fraden. New York: American Institute of Physics, 1996.

Authoritative handbook on every aspect of modern sensors.

Handbook of Optical Character Recognition and Document Image Analysis. P.S.P. Wang, H. Bunke. New York: World Scientific Pub. Co., 1997.

A technical handbook on optical character recognition.

Integrated Optics: Theory and Technology. Robert G. Hunsperger. New York: Springer-Verlag, 1995.

A comprehensive guide to optical integrated circuits, including their relationship to fiber optics.

Practical Holography. Graham Saxby. New York: Prentice Hall, 1994.

A practical and easy-to-understand treatment of holography.

Principles of Transistor Circuits: Introduction to the Design of Amplifiers, Receivers, and Digital Circuits. S.W. Amos. Newton, MA: Butterworth-Heinemann, 1994.

A technical overview of transistor circuits.

Satellite Communications Pocket Book. James Wood. Newton, MA: Butterworth-Heinemann, 1994.

The past, present, and future technologies of satellite communications.

Articles

"Cook up a Kitchen on a Computer." S. Hertz. *House Beautiful,* September 1996, 78.

How CAD/CAM is used to design a modern home.

"Fit for a King (Computer Scanners Capture the Body's Subtle Shapes for Custom Fit Clothing)." R. Lipkin. *Science News,* May 18, 1996, 316-17.

How computer CAD/CAM is used in the manufacture of clothing.

"First Light for Integrated Light Emitter." C. Seife. *Science,* November 29, 1996, 1465.

New technologies in integrated optics.

"Flat Panels Invade the Desktop." C. Lu. *MacWorld* September 1996, 43.

The latest in flat panel CRTs.

"Iridium Marks Dawn of New Industry." J. C. Anselmo. *Aviation Week & Space Technology,* January 13, 1997, 363-64.

The emerging communication satellite technology, and the possible future.

"Monitor Your TV." *Popular Electronics,* September 1996, 27.

The current development in integrated multimedia monitors.

"The Next Wave: DMD, PDP, and LCD Technology are Poised to Remake the TV Landscape." T. Heald. *Video,* September 1996, 32-38.

Upcoming technologies for television screens.

"1997: A New Space Odyssey." W.J. Cook. *U.S. News & World Report,* March 3, 1997, 44-48.

A synopsis of the new communications satellites.

"Painting Pictures with Atom Waves." J. Glanz. *Science,* August 9, 1996, 737.

A look at holography.

"Plastic Transistors Gain Speed on Silicon." R.F. Service. *Science,* August 16, 1996, 879.

The effectiveness of new types of transistors.

"Tough Stuff." W.W. Gibbs. *Scientific American,* March 1996, 34-35.

A close look at modern ceramic composites.

Index

by Irv Hershman

Boldface page numbers show where primary definitions will be found.